Plant–Plant Allelopathic Interactions

Udo Blum

Plant–Plant Allelopathic Interactions

Phenolic Acids, Cover Crops and Weed Emergence

Springer

Udo Blum
Department of Plant Biology
North Carolina State University
Raleigh, NC 27695-7612
USA
udo_blum@ncsu.edu

ISBN 978-94-007-0682-8 e-ISBN 978-94-007-0683-5
DOI 10.1007/978-94-007-0683-5
Springer Dordrecht Heidelberg London New York

Library of Congress Control Number: 2011922311

Printed on acid-free paper

Springer is part of Springer Science+Business Media (www.springer.com)

This book is dedicated to all who have labored and will labor in the field of plant–plant allelopathic interactions.

Preface

For part of my PhD thesis I characterized the distribution of tannic acids in soils underneath sumac (*Rhus copallina* L.) located in abandoned fields of central Oklahoma (Blum and Rice 1969). Large quantities of tannic acids were found in the litter and organic residues underneath sumac. Tannic acids, which are very water soluble, were also found in the soil to a depth of 75 cm, with a definite zone of concentration at 45–55 cm. The techniques utilized at the time to recover and quantify tannic acids were rudimentary, at best. Amounts below 400 ppm added to soils could not be recovered, even though concentrations as low as 33 ppm added to soils inhibited nodulation of red kidney beans (*Phaseolus vulgaris* L. "Burpee"). These observations and their implications to plant–plant allelopathic interactions intrigued me at the time and I made a promise to myself that I would take another look at this subject in the future. Around 1980 I was ready to fulfill that promise. For the next 20 plus years research in my laboratory was primarily focused on various aspects of plant–plant allelopathic interactions with an emphasis on seedling behavior, soil chemistry, and microbiology. This book is a summary and retrospective analysis of this research program.

Although research publications on allelopathy have increased at a phenomenal rate since the 1980s, what is generally lacking are in-depth analyses and integration of this literature. For example, a quick search of Science Citation Index yielded 112 publications between 1981 and 1990, 627 publications between 1991 and 2000, and 1,615 publications between 2001 and 2010. The terms "allelopathic interactions" yielded 6, 58, and 212 publications over the same time intervals. However, less than 10% of these 276 citations listed for allelopathic interactions could be classified as review papers for allelopathic interactions of higher plants. These reviews, with minor exceptions, summarized, described, pooled, and/or integrated data for plant–plant allelopathic interactions determined for different species, environments, and ecosystems utilizing a range of different methods/protocols. Such reviews are useful in that they can identify potential/likely mechanisms that may bring about plant–plant allelopathic interactions and provide general guidelines and directions for future research. However, to identify and determine actual mechanisms that control and/or regulate the expression of plant–plant allelopathic interactions within a given ecosystem requires in-depth quantitative analyses of individual ecosystem processes and their interactions utilizing consistent experimental protocols. The research described in this book is an attempt to do just that for one type of ecosystem.

This book does not provide a comprehensive review of the plant–plant allelopathic interaction literature. For a general review of this literature the reader may wish to read several of the following: Rice (1974, 1979, 1983, 1984, 1995), Putnam and Tang (1986), Waller (1987), Siqueria et al. (1991), Inderjit et al. (1995, 1999), Inderjit and Keating (1999), Macías et al. (1999, 2004), Reigosa et al. (2006), Fujii and Hiradate (2007), Willis (2007), and Zeng et al. (2008).

There are several things that are unique about this book:

a. The general format is that of research papers published in scientific journals. The materials are organized in sections such as, Abstract, Introduction, Materials and Methods, and Results and Discussion.
b. There are four chapters, including an introduction to allelopathic plant–plant interactions (Chapter 1). They all emphasize basic aspects of science, but Chapter 2 is more theoretical/hypothetical in nature, Chapter 3 is more practical in nature, and Chapter 4 integrates the information presented in Chapters 2 and 3 and suggests future direction for research in plant–plant allelopathic interactions.
c. Comments regarding logic, reasons, and justifications, for various procedures used are provided throughout the book.
d. The Scientific Method and its approach to research are emphasized. For example, instead of definitive conclusions at the end of the book cons and pros are provided so that readers can draw their own conclusions. The reader will also find an extended listing of if-then hypotheses, and
e. Although a broad range of literature is included, the primary focus of this book is a summary and retrospective analysis of some 20 plus years of research on plant–plant allelopathic interactions at North Carolina State University.

The above format was chosen so that researchers, students, farmers, as well as layman interested in science, reduced tillage production, and plant–plant allelopathic interactions, in particular, can learn to appreciate and understand the nature of science, its benefits and limitations, and our present knowledge of the action of natural products such as phenolic acids in soil on plant growth and development.

Raleigh, NC Udo Blum
August 19, 2010

References

Blum U, Rice EL (1969) Inhibition of symbiotic nitrogen-fixation by gallic and tannic acid and possible roles in old-field succession. Torrey Bot Club 96:531–544

Fujii Y, Hiradate S (2007) Allelopathy: new concepts and methodology. Science Publishers, Enfield, NY

Inderjit, Keating KI (1999) Allelopathy: principles, procedures, processes, and promises for biological control. Adv Agro 67:141–231

Inderjit, Daskshini KMM, Einhellig FA (1995) Allelopathy: organisms, processes, and applications. ACS symposium series, vol 582. American Chemical Society, Washington, DC
Inderjit, Daskshini KMM, Foy CL (1999) Principles and practices in plant ecology: allelochemical interactions. CRC Press, Boca Raton, FL
Macías FA, Galindo JGC, Molinillo JMG, Cutler H (1999) Recent advances in allelopathy I. A science for the future. Cádiz University Press, Puerto Real Cádiz, Spain
Macías FA, Galindo JGC, Molinillo JMG, Cutler H (2004) Allelopathy: chemistry & modes of action of allelochemicals. CRC Press, Boca Raton, FL
Putnam AR, Tang CS (1986) Science of allelopathy. Wiley, New York, NY
Reigosa MJ, Pedrol N, Gonzalez L (2006) Allelopathy. A physiological process with ecological implications. Springer, Dordrecht, The Netherlands
Rice EL (1974) Allelopathy. Academic Press, Orlando, FL
Rice EL (1979) Allelopathy – an update. Bot Rev 45:15–109
Rice EL (1983) Pest control with nature's chemicals: allelochemics and pheromones in gardening and agriculture. University of Oklahoma Press, Norman, NY
Rice EL (1984) Allelopathy. Academic Press, Orlando, FL
Rice EL (1995) Biological control of weeds and plant diseases: advances in applied allelopathy. University of Oklahoma Press, Norman, NY
Siqueira JO, Nair MG, Hammerschmidt R, Safir GR (1991) Significance of phenolic compounds in plant-soil-microbial systems. Crit Rev Plant Sci 10:63–121
Waller GR (1987) Allelochemicals: role in agriculture and forestry. ACS symposium series, vol 330. American Chemical Society, Washington, DC
Willis RJ (2007) The history of allelopathy. Springer, Dordrecht, The Netherlands
Zeng RS, Mallik AU, Luo SM (2008) Allelopathy in sustainable agriculture and forestry. Springer, New York, NY

Acknowledgements

Although my research interests in allelopathy have been a primary focus for most of my academic career, I did take several excursions into other research areas (e.g., air pollution biology, and salt marsh ecology) before returning full time to the subject matter of allelopathy. In retrospect these excursion turned out to be extremely beneficial to my understanding of stress physiology and ecosystem biology, important insights needed when studying plant–plant allelopathic interactions. My teaching of beginning and advanced undergraduate botany courses and graduate courses in plant physiology, ecology, plant physiological ecology, and root ecology also proved to be invaluable in my pursuit of understanding the mechanisms of plant–plant allelopathic interactions by providing me with an opportunity to develop a much more in-depth appreciation of plant morphology, anatomy, physiology, and population biology, and soil physics, chemistry and microbiology.

Equally as important as a solid understanding of plant, microbial, and soil biology was an appreciation of the scientific method. The importance of the scientific method as a tool for studying biological systems was instilled within me by EL Rice, my PhD mentor at The University of Oklahoma, and was reinforced by my teaching of botany courses using the Socratic Method at both the University of Oklahoma and at North Carolina State University.

I also want to acknowledge the help of several statisticians at North Carolina State University who over the years provided me with the opportunity to develop and refine my skills in experimental design, data analysis, and modeling. In particular, I would like to express my appreciation to Professors RJ Monroe, JO Rawlings, and TM Gerig of the Department of Statistics.

Along the way there were numerous faculty members, graduate and undergraduate students, and technicians who influenced, shaped, and reshaped my research program in allelopathy. A deep felt thank you to all of them. In particular, I would like to express my appreciation to faculty members C Brownie, RC Fites, TM Gerig, F Louws, LD King, SR Shafer, SB Weed, TR Wentworth, and AD Worsham, visiting scientist S-W Lyu, technicians/graduate students BR Dalton and K Klein, graduate students MF Austin, CL Bergmark, FL Booker, LJ Flint, AB Hall, LD Holappa, M Kochhar, ME Lehman, JV Perino, KJ Pue, J Rebbeck, JR Shann, K Staman, ER Waters, and AG White, and the assistance of CG Van Dyke in processing the samples and taking the electron micrographs of microbial populations on cucumber root surfaces.

I would also like to acknowledge the following organizations for providing research support and/or funding: North Carolina Agricultural Research Service, USDA Competitive Research Grants Program, Southern Region Low-Input Agricultural Systems Research and Extension Program, North Carolina Agricultural Foundation Graduate Research Assistantship Program, and the Departments of Botany (now Plant Biology), Soil Science, and Statistics.

Finally, the author wishes to thank MA Blum, SO Duke, JR Troyer, JD Weidenhamer, and AD Worsham for editing, reviewing, and for thoughtful and constructive comments.

Contents

Abbreviations

ACT	Basal medium for actinomycetes
CAF	Caffeic acid
C-clover	Crimson clover
CFU	Colony-forming units
C/N	Carbon/nitrogen ratio
DBW3	EDTA extraction of soil at room temperature and soil extraction ratio of 1:100
DTPA	Diethylenetriaminepentaacetic acid
EDTA	Ethylenediaminetetraacetic acid
FER	Ferulic acid
GUE	Sodium hydroxide extraction of soil at room temperature and soil extraction ratio of 1:1 (GUE2) or at 121°C and soil extraction ratio of 1:43 (GUEN)
GLM	General linear model
GLU	Glucose
HPLC	High performance liquid chromatograph
kv	kilovolts
MEOH	Methanol
MES	2-(N-morpholino) ethanesulfonic acid
mOsm	milliosmoles
NLIN	Non linear
OMe	Methoxy
PEG	Polyethylene glycol
PPFD	Photosynthetic photon flux density
PCO	*p*-Coumaric acid
PDMS	Polydimethylsiloxane
POH	*p*-Hydroxybenzoic acid
PRO	Protocatechuic acid
PVP	Polyvinylporrolidone
RCM-100	Radical Pak cartridge
R	Root
R + S	Root plus shoot

S	Shoot
S-clover	Subterranean clover
SIN	Sinapic acid
SYR	Syringic acid
VAN	Vanillic acid

List of Figures

List of Tables

Chapter 1
Plant–Plant Allelopathic Interactions

Abstract This chapter provides a definition of allelopathy, processes included and excluded from plant–plant allelopathic interactions, criteria needed to establish plant–plant allelopathic interactions, application of the Scientific Method, use of laboratory and field bioassays, and a short paragraph on the subject matter of the rest of the book.

Molisch (1937) coined the term allelopathy to describe both positive and negative chemical interactions of plants mediated through the environment (Willis 1985, 2007; Molisch [translation] 2001). His definitions of allelopathy and allelopathic agents as translated by LJ La Fleur and MAB Mallik (2001) were as follows, respectively:

> The already described phenomenon of the effects of one plant on the other, which in physiology plays and will play an important role, deserves to be described by a special term. I propose for it the word 'Allelopathy', derived from two Greek Words: ... (allelon) means reciprocal, mutual, among each other, and ... (pathos) means grief, sorrow, or that which happens to one.
>
> Inhibitory material or substance which is released from one plant and exerts an influence on another plant, ... It deals with an influence which is exhibited in a remarkable way in spatially separated plants. (Quotations used with permission of International Allelopathy Foundation)

Although not explicitly stated by him, the allelopathic substances alluded to in his book were organic compounds, e.g., ethylene (Molisch 1937, [translation] 2001).

While the definition of allelopathy has varied over time (see Grodzinsky 1971; Whittaker and Feeny 1971; Rice 1974, 1979, 1983, 1984; Willis 1985, 1994, 2007), Molisch's definition is now the central core of the definition used by most researchers studying plant–plant allelopathic interactions. The inclusion of "chemical interactions mediated between spatially separated plants" in his definition excludes from allelopathic interactions:

a. the effects of physicochemical modifications of the soil (other than promoters and inhibitors) resulting from the addition of organic substances to the soil,

U. Blum, *Plant–Plant Allelopathic Interactions*, DOI 10.1007/978-94-007-0683-5_1,

b. the utilization of organic substances as a source of nutrients, carbon, and/or energy (Whittaker and Feeny 1971), and
c. the direct transfer of organic substances between plants through mycorrhizae and/or root grafts (Blum 2006).

The inclusion of c. may be questioned by some since transfer by mycorrhizae is never direct. Organic (as well as inorganic) substances have to pass through intercellular spaces or the apoplast while being transferred from or into plant root cells by way of mycorrhizae. Furthermore, even with root grafts there is a transfer from one plant to another, although in this case the transfer is direct. As presently used in the literature, plant–plant allelopathic interactions are due to the action of organic substances released from living plants or decomposing recognizable plant tissues/residues into the environment where these may be oxidized or reduced before influencing in a positive or negative manner associated plants. The positive or negative influences on plants can either be due to direct (e.g., impact on root membranes and/or cell process) or indirect (e.g., impact on nodule or mycorrhizae formation, development, and/or function) actions of allelopathic agents.

Plant–plant allelopathic (or chemical) interactions are extremely complex and identifying and characterizing such interactions in nature are difficult. Of course, the same could be said for plant–plant resource competition. Both plant–plant resource competition and plant–plant allelopathic interactions are in a sense intuitive. Plants require resources such as space, energy, water, and nutrients and these resources are frequently limiting in nature. Plants are chemical factories that are surprisingly leaky to organic substances (as well as inorganic substances) that may be stimulatory, neutral, or inhibitory to biological processes of associated plants. While plant–plant resource competition has been readily accepted by plant biologists and ecologists, the same has not been the case for plant–plant allelopathic interactions, even though chemical interactions (i.e., allelopathy) between insects and plants, microbes and plants, microbes and microbes, and animals and plants have been readily accepted. The difference in acceptance between resource competition and plant–plant allelopathic interactions was likely due to the following:

a. the definition of allelopathy has been altered several times,
b. the bioassays of early plant–plant allelopathic interaction studies were frequently poorly designed,
c. the forceful skepticism of opponents to plant–plant allelopathic interactions, and
d. the rigor of proof required for plant–plant allelopathic interactions was set to a higher standard than those for resource competition (Harper 1975, 1977; Stowe 1979; Williamson 1990; Romeo 2000; Willis 2007).

The justification for the difference in standards of proof is, however, not entirely clear. It may have been a response to b. and c. above and/or may have occurred because the concept of plant–plant allelopathic interactions challenged the long held paradigm of resource competition. In addition, organic substances involved in plant–plant allelopathic interactions can be exceedingly ephemeral, i.e., are readily metabolized, leached, volatilized, and/or bound (sorbed). Interestingly enough, similar behavior can be observed for resources such as energy, water, and nutrients (e.g., nitrogen, potassium, and phosphorus).

The challenge, therefore, has been and continues to be to establish realistic and workable criteria that can be used to provide an acceptable level of proof for plant–plant allelopathic interactions (Willis 1985; Williamson 1990; Romeo and Weidenhamer 1998; Blum et al. 1999). In general the following criteria have now been adopted:

a. Patterns of stimulation or inhibition of plants on other plants must be shown.
b. The observed patterns cannot be solely explained by physicochemical modifications of the environment (other than promoters and/or inhibitors), utilization of substances as a source of nutrients, carbon and/or energy, transfer through mycorrhizae and/or root grafts (please note ambiguity above), and/or biotic factors such as resource competition, herbivory, or disease.
c. The putative allelopathic plants or their residues must produce/contain and release organic substances into the environment that will ultimately be capable of stimulating or inhibiting the function or growth of associated plants.
d. The affected plants must come in contact with and interact with the organic promoters or inhibitors produced directly or indirectly (e.g., modified by microorganisms) by an allelopathic plant, and
e. These organic promoters and inhibitors must be at sufficient concentrations and be present for sufficient length of time to modify plant function and growth of receiving plants either directly (e.g., impact on root membranes and/or cell process) or indirectly (e.g., impact on nodule or mycorrhizae formation, development, and/or function).

Demonstrating all of these criteria in nature turns out to be quite a challenge. Patterns of stimulation or inhibition (criterion a.) can be readily determined. However, meeting the other criteria (criteria b. to e.) is much more difficult. For example, Connell (1990) describes how indirect interactions via shared enemies such as herbivores or disease or indirect interactions via other species on the same trophic level might lead to the conclusion of resource competition between two species when in reality that is not the case. Plant–plant allelopathic interactions could also be misidentified as resource competition. Separating the action of resource competition from plant–plant allelopathic interactions appears even more challenging than identifying shared enemies, particularly when competition and allelopathic interactions occur simultaneously. Inderjit and Del Moral (1997) would argue that plant–plant allelopathic interactions and resource competition are so interrelated that they cannot be separated in natural systems. Others would argue that under the right experimental conditions density-dependent inhibition may be used to distinguished plant–plant allelopathic interactions from resource competition (Weidenhamer et al. 1989; Thijs et al. 1994; Weidenhamer 1996) as long as autotoxicity and/or hormesis are absent (An et al. 1993; Belz et al. 2005; Sinkkonen 2001, 2003, 2007). The use of activated carbon to bind allelopathic agents may also be helpful in separating plant–plant allelopathic interactions and resource competition

(Mahall and Callaway 1992; Nilsson 1994; Wu et al. 2000; Ridenour and Callaway 2001; Bertin et al. 2009). When one adds that:

a. the production and release from plants, modifications once released into the environment, and subsequent actions and/or functions of allelopathic agents vary with the physicochemical and biotic environment;
b. organic compounds can be stimulatory, neutral, or inhibitory depending on their physicochemical states, concentrations, and environments;
c. many organic compounds thought to be involved in plant–plant allelopathic interactions appear to have a short half life or a rapid turnover rate;
d. sensitivity of plant taxa and plant processes are highly variable and change with age, state of acclimation, plant "health", and physicochemical and biotic environment; and
e. soil solutions, soil processes, and plant growth and development are very complex and variable over time; then the difficulties of identifying and using appropriate and precise methodologies to characterize plant–plant allelopathic interactions and separating or isolating them from other real and/or apparent interactions become all too obvious.

In a previous publication I stated that:

> For the science of allelopathy, like all other sciences, there is a prescribed methodology by which problems are to be addressed and solved, the "Scientific Method". Once a problem has been identified, this method requires that alternative hypotheses, tentative explanations, be generated which can be experimentally tested. Occurrence (acceptance) or non-occurrence (rejection) of predictions deduced for each hypothesis is then determined by means of experiments. Finally, science progresses not by trying to confirm hypotheses but by attempting to falsify them since it is usually possible to find at least some confirmatory evidence for any hypothesis, but one solid piece of negative data refutes a hypothesis completely (Blum 2007; Quotation used with permission of Science Publishers).

I would be remiss here if I did not mention that some would argue that the method described above is an over simplification of what really happens when successful scientists do science. In fact, the truth is there are many ways to do science. Scientists should feel free to use any method that is appropriate, to use their imagination and intuition, and if necessary break the formulated rules (Laidler 1998). Obviously there are limits. For example, data should never be fabricated, but sometimes it is justifiable to ignore some data particularly when they appear to be due to extraneous factors. In the process mistakes may be made but ultimately, even with such mistakes, the progress of science does not suffer. For specific examples see Laidler (1998).

In biology the most frequently used tools for experimentally testing alternative hypotheses are field and/or laboratory bioassays. These tools, however, have their limits in linking cause with effect. For example:

a. In absolute terms designing field experiments (bioassays) and including all the necessary controls (assuming we even knew all the required controls) to link cause with effect is presently not possible.

b. "Teasing out" direct or indirect linkages between active environmental factors (e.g., colinearities) and thus cause and effect are extremely difficult under field conditions, even under the best of circumstances. In fact, this may be impossible with any certainty, and
c. Determining the actual (primary) forces acting on plants at a given point in time under field conditions are essentially impossible because of the dynamics of the environment, colinearities of environmental factors, and time delays between cause and "visible" or detectable effects.

Laboratory bioassays have similar problems to those of field bioassays, but on a smaller scale, i.e., they are smaller "black boxes" (Blum 1999, 2007). However, the physical, chemical and biological characteristics of laboratory bioassays can be more readily manipulated to hold some characteristics constant while others are being modified. The benefits of this are obvious for characterizing cause and effect relationships, but these benefits come at a cost. The cost:

a. we create artificial environments that may never occur in the field;
b. we modify or eliminate feedback mechanisms that operate in the field; and
c. the low level of scale limits our ability to generalize about the observed cause and effect relationships.

The bottom line is that both field and laboratory bioassays are essential when attempting to identify plant–plant allelopathic interactions in the field. Field observations and bioassays provide the basis and direction of research (e.g., the formation of hypotheses), while laboratory experiments help to identify potential mechanisms operating under field conditions. Ultimately, however, no matter how sophisticated our observation and bioassays, the final conclusions will always have to be based on common sense, since with few exceptions, our conclusions are based on partial information. Our goal may be to achieve Truth with a capital T but at any point in time we are working with small t truth. Nature is very complex!

What follows this introduction to plant–plant interactions (Chapter 1) are three additional chapters. The first chapter (Chapter 2) describes the behavior of allelopathic agents in nutrient culture and soil-microbe-seedling systems under laboratory conditions. Simple phenolic acids were chosen as the allelopathic agents for study in these model systems (see justifications in Section 2.2.6). The next chapter (Chapter 3) describes the relationships or lack of relationships between weed seedling behavior and the physicochemical environment in cover crop no-till fields and in laboratory bioassays. Here as well the emphasis is on the potential role of phenolic acids. The final chapter (Chapter 4) restates the central objectives of Chapters 2 and 3 in the form of testable hypotheses, addresses several central questions raised in these chapters, outlines why a "holistic" approach is required when studying allelopathic plant–plant interactions, and suggests some ways by which this may be achieved.

References

An M, Johnson IR, Lovett J (1993) Mathematical modeling of allelopathy: biological response to allelochemicals and its interpretation. J Chem Ecol 19:2379–2388

Belz RG, Hurle K, Duke SO (2005) Dose response – a challenge for allelopathy? Nonlin Biol Toxicol Med 3:173–211

Bertin C, Harmon R, Akaogi M, Weidenhamer JD, Weston LA (2009) Assessment of the phytotoxic potential of *m*-tyrosine in laboratory soil bioassays. J Chem Ecol 35: 1288–1294

Blum U (1999) Designing laboratory plant debris-soil bioassays: some reflections. In: Inderjit, Daskshini KMM, Foy CL (eds) Principles and practices in plant ecology: allelochemical interactions. CRC Press, Boca Raton, FL, pp 17–23

Blum U (2006) Allelopathy: a soil system perspective. In: Reigosa MJ, Pedrol N, González L (eds) Allelopathy: a physiological process with ecological implications. Springer, Dordrecht, The Netherlands, pp 299–340

Blum U (2007) Can data derived from field and laboratory bioassays establish the existence of allelopathic interactions in nature? In: Fujii Y, Hiradate S (eds) Allelopathy: new concepts and methodologies. Science Publishers, Enfield, NH, pp 31–38

Blum U, Shafer SR, Lehman ME (1999) Evidence for inhibitory allelopathic interactions involving phenolic acids in field soils: concepts vs. an experimental model. Crit Rev Plant Sci 18:673–693

Connell JH (1990) Apparent versus "real" competition in plants. In: Grace JB, Tilman D (eds) Perspectives of plant competition. Academic Press, New York, NY, pp 9–26

Grodzinsky AM (1971) Problems and results of allelopathy in the work of Soviet scientists. In: National Academy of Sciences (ed) Biochemical interactions among plants. National Academy of Sciences, Washington, DC, pp 44–51

Harper JL (1975) Allelopathy (a review). Q Rev Biol 50:493–495

Harper JL (1977) Population biology of plants. Academic Press, New York, NY

Inderjit, Del Moral R (1997) Is separating resource competition from allelopathy realistic? Bot Rev 63:221–230

Laidler KJ (1998) To light such a candle chapters in the history of science and technology. Oxford University Press, Oxford

Mahall BE, Callaway RM (1992) Root communication mechanisms and intracommunity distribution of two Mojave desert shrubs. Ecology 73:2145–2151

Molisch H (1937) Der Einfluss einer Pflanze auf die andere – allelopathie. Fisher, Jena

Molisch H (2001) The influence of one plant on another: allelopathy. In: Narwal SS (ed) LaFleur LJ and Mallik MAB (trans: from German). Scientific Publishers, Jodhpur, India

Nilsson MC (1994) Separation of allelopathy and resource competition by the boreal dwarf shrub *Empetrum hermaphroditum* hagerup. Oecologia 98:1–7

Rice EL (1974) Allelopathy. Academic Press, Orlando, FL

Rice EL (1979) Allelopathy An update. Bot Rev 45:15–109

Rice EL (1983) Pest control with nature's chemicals: allelochemics and pheromones in gardening and agriculture. University of Oklahoma Press, Norman, OK

Rice EL (1984) Allelopathy. Academic Press, Orlando, FL

Ridenour WM, Callaway RM (2001) The relative importance of allelopathy in interference: the effects of an invasive weed on a native bunchgrass. Oecologia 126:444–450

Romeo JT (2000) Raising the beam: moving beyond phytotoxicity. J Chem Ecol 26:2001–2014

Romeo JT, Weidenhamer JD (1998) Bioassays for allelopathy in terrestrial plants. In: Haynes KF, Millar JG (eds) Methods in chemical ecology: bioassay methods. Chapman & Hall, New York, NY, pp 179–211

Sinkkonen A (2001) Density-dependent chemical interference – an extension of the biological response model. J Chem Ecol 27:1513–1523

Sinkkonen A (2003) A model describing chemical interference caused by decomposing residues at different densities of growing plants. Plant Soil 250:315–322

Sinkkonen A (2007) Modeling the effects of autotoxicity on density-dependent phytotoxicity. J Theor Biol 244:218–227

Stowe LG (1979) Allelopathy and its influence on the distribution of plants in an Illinois old-field. J Ecol 67:1065–1085

Thijs H, Shann JR, Weidenhamer JD (1994) The effects of phytotoxins on competitive outcome in a model system. Ecology 75:1959–1964

Weidenhamer JD (1996) Distinguishing resource competition and chemical interference: overcoming the methodological impasse. Agron J 88:866–875

Weidenhamer JD, Hartnett DC, Romeo JT (1989) Density-dependent phytotoxicity: distinguishing resource competition and allelopathic interference in plants. J Appl Ecol 26:613–624

Whittaker RH, Feeny PP (1971) Allelochemics: chemical interactions between species. Science 171:757–770

Williamson GB (1990) Allelopathy, Koch's postulates, and the neck riddle. In: Grace JB, Tilman GD (eds) Perspectives in plant competition. Academic Press, San Diego, CA, pp 143–162

Willis RJ (1985) The historical bases of the concept of allelopathy. J Hist Biol 18:71–102

Willis RJ (1994) Terminology and trends in allelopathy. Allelopathy J 1:6–28

Willis RJ (2007) The history of allelopathy. Springer, Dordrecht, The Netherlands

Wu H, Pratley J, Lemerle D, Haig T (2000) Laboratory screening for allelopathic potential of wheat (*Triticum aestivum*) accessions against annual ryegrass (*Lolium rigidum*). Aust J Agric Res 51:259–266

Chapter 2
Plant–Plant Allelopathic Interactions. Phase I: The Laboratory

Abstract This chapter describes the underlying criteria and assumptions in the development and execution of bioassays utilizing model systems in the laboratory. It provides details and commentary regarding the materials and methods used. Describes and discusses the results and significance of the following: (a) effects and duration of effects for individual phenolic acids, mixtures of phenolic acids, and phenolic acids in combination with other organic and inorganic compounds on seedlings growing in nutrient and in soil cultures, (b) the interactions of phenolic acids with seedling roots, bulk-soil and rhizosphere microbes, and soil particles, (c) the relationships between phenolic acid-utilizing microbes, soil characteristics, and seedling inhibition, (d) the effects of phenolic acids on the various life stages of plants, (e) what happens to phenolic acid concentrations over time in nutrient culture and seedling-microbe-soil systems, and (f) the relevance of laboratory bioassays to field studies.

By the middle of the nineteen eighties there was already an extensive literature on:

a. plant–plant allelopathic interactions,
b. the identification, extraction, and phytotoxicity of various potential allelopathic agents in soil and plant materials (e.g., phenolic acids),
c. the environmental and biotic factors affecting the production of allelopathic agents in plants,
d. the release of potential allelopathic agents into the soil from living and dead plant materials (e.g., root exudates, secretions and lysates, leaf leachates, and decomposition and leaching of plant residues),
e. the sorption and leaching of potential allelopathic agents in the soil,
f. the production and loss of allelopathic agents in the soil by the action of soil microbes, and
g. their mechanisms and modes of action (see Rice 1974, 1979, 1983, 1984, 1986; Thompson 1985; Henry 1985, 1987; Putnam and Tang 1986; Waller 1987).

The conventional wisdom of the time was that the inhibition by water soluble allelopathic agents was determined by the sensitivity of the plant species, the type of allelopathic agent or agents present in the soil and their net concentration in the

U. Blum, *Plant–Plant Allelopathic Interactions*, DOI 10.1007/978-94-007-0683-5_2,

soil over time (input-output). An early appreciation of how allelopathic plant–plant interactions might be modified by the physicochemical environment (e.g., soil pH, mineral nutrition, temperature, and drought) was also evident in the literature (Hall et al. 1982; Rice 1984; Balke 1985; Blum et al. 1985b, 1987; Einhellig 1987). However, a coherent understanding of what happens to inhibitory allelopathic agents once they enter the soil environment and how this may influence the expression of plant–plant allelopathic interactions was still in its infancy. This lack of understanding lead us to develop laboratory model systems to study how plant and soil processes (Fig. 2.1) might influence plant–plant allelopathic interactions. The focus was on inhibitory plant–plant allelopathic interactions since only limited data on stimulatory allelopathic plant–plant interactions was and continues to be available in the literature (Rice 1986; Pandey 1994; Duke et al. 2006; Belz et al. 2007; Belz 2008).

2.1 Criteria for Model Systems

We established the following criteria for our model systems:

a. The plant bioassay species must germinate rapidly and consistently, grow rapidly in nutrient culture and in small soil containers under light banks, have consistent leaf characteristics that will allow determination of nondestructive leaf area expansion (i.e., growth), and be susceptible and respond in a consistent manner to the allelopathic agents being tested.
b. The soil substrates must have chemical and physical characteristics for seedling growth, amenable to the recovery of root systems, maintain functional microbial populations even after soils have been air dried and stored, and represent a range of soil types found in the Piedmont and Coastal Plain of the Eastern United States.
c. The allelopathic agents must have been identified as potential inhibitors, are readily available in pure form, have a range of derivatives (e.g., different side chains and side groups) that vary in inhibition, are soluble in water at reasonable rates, are recoverable by a range of extraction procedures from soils and plant materials, and are readily identifiable and quantifiable once recovered.
d. Experimental units (e.g., petri dishes, nutrient culture bottles, and soil containers) must be small enough to:

    1. minimize seed or root environmental variation,
    2. maximize rapid and consistent seed or root contact when inhibitors are added to the experimental units, and
    3. minimize the number and size of subsamples required for reasonable estimates of inhibitor concentrations and microbial populations in the medium or in contact with roots, and must be large enough to minimize the daily frequency of irrigation and addition of mineral nutrients, and

e. Environmental conditions must be reasonably controllable and repeatable over time.

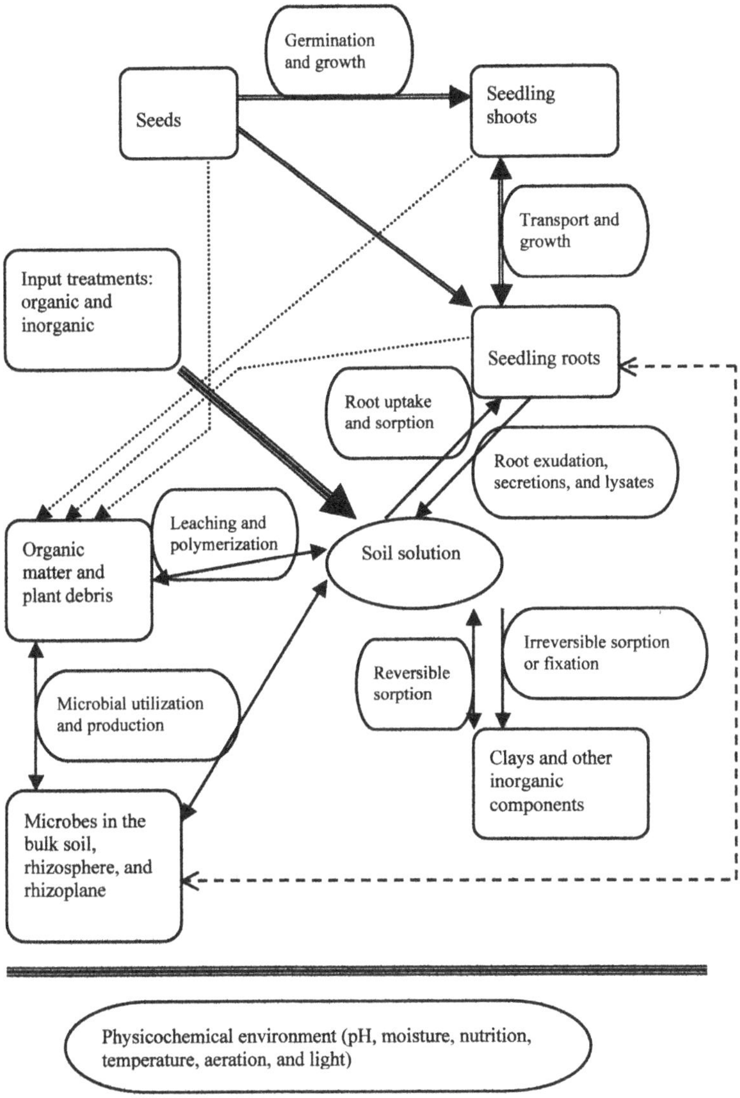

Fig. 2.1 A seedling-microbe-soil model system

2.2 Materials, Methods, and Commentary

It is not uncommon to find in the research literature that experimental procedures (e.g., dose response studies) are taken from previous publications and repeated with slight modification (e. g., different species and/or environments). That may be appropriate if the primary goal is simply to describe similarities or differences in behavior from previous studies. However, if the goal is to understand mechanisms of action, this approach is problematic. To determine mechanisms of action requires an in-depth understanding of the behavior of the individual components as well as the interactions of those components making up a system under study. Unfortunately this behavior varies with components and combinations of components making up a system. Thus in-depth understanding of such behavior is lacking in most instances because obtaining the baseline data required is costly both in terms of resources and time. From the beginning, however, we made a commitment to collect such baseline data wherever and whenever possible.

2.2.1 General Bioassay Procedures

Germination and radicle/hypocotyl/epicotyl growth studies were carried out in petri dishes (100 × 15 mm) containing seeds, No. 3 Whatman filter paper, and 5–7 ml of a pH adjusted test solution (pH 4.5–7.5), with or without MES buffer (2-[N-morpholino] ethanesulfonic acid), in the dark at 25–30°C (Blum et al. 1984). Soil extracts or solutions containing individual phenolic acids or phenolic acid mixtures were tested. Seed numbers in the petri dishes varied with seed size, but were held constant for a given species. Maintaining a constant solution volume and seed number in petri dishes is important since changes in solution volume or seed number can modify the amount of inhibition observed for a given concentration of inhibitor (Weidenhamer et al. 1987). For example, a larger solution volume or a lower seed number per petri dish can increase the level of inhibition observed. Times for measurements of radicle/hypocotyl/epicotyl lengths varied with seed species. Hereafter radicle/hypocotyl/epicotyl lengths will simply be referred to as radicle/hypocotyl lengths since epicotyl lengths were absent or extremely small. In a number of experiments sterile techniques were employed in setting up the germination experiments. For those all solutions were filter-sterilized (0.2-μm membrane filter) and all other materials were autoclaved. However, since concentrations of sodium hypochloride required to consistently surface sterilize cucumber seeds (our primary bioassay species; see Section 2.2.2) influenced radicle/hypocotyl growth, seed surfaces were not sterilized (Blum et al. 1984).

For seedling experiments, seeds were germinated in the dark at 25–30°C in vermiculite moistened with water or Hoagland's nutrient solution (Hoagland and Arnon 1950). After 48 h seedlings were transferred to snap-cap bottles (Blum and 1985), soil cups (Blum et al. 1987), or soil columns (Blum et al. 1999a). Seedlings for the split-root system were grown an additional 8–10 days in snap-cap bottles before being transferred to split-root containers (Klein and Blum 1990a; Lyu and Blum 1990; Lehman et al. 1994). Seedlings were grown under a series of light banks composed of 25-W incandescent bulbs and 40-W cool white fluorescent tubes (Fig. 2.2). Light banks provided approximately 140–180 $\mu mol/m^2/s$ photosynthetic photon flux density (PPFD) from 8 AM to 8 PM at the seedling level. The roots of the seedlings were protected from light. The temperature in the light banks ranged from 21° to 30°C. Experimental units were arranged in a random or randomized-block design within the light banks.

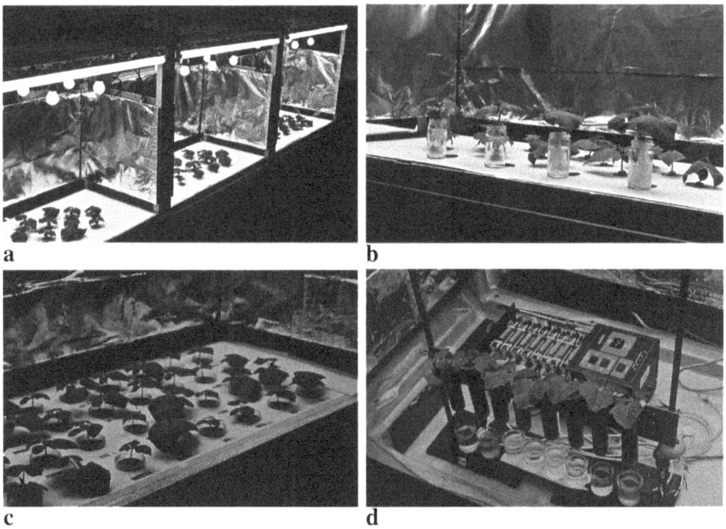

Fig. 2.2 Light banks: **a** general view, **b** nutrient culture, **c** soil cup system, and **d** continuous-flow system

For nutrient culture (snap-cap bottles and split-root containers), seedlings were supplied with new nutrient solution (Hoagland and Arnon 1950) every other day (controls) or new nutrient solutions containing individual phenolic acids or phenolic acid mixtures (treatments). Solutions were changed in the morning. Water was added once in the afternoon, as needed, on the day when solutions were changed or in the morning and afternoon, as needed, on days when solutions were not changed. Total volume in each bottle was brought to a constant level each time solutions were changed or water was added. For the soil cup systems nutrient solution plus water or phenolic acid treatment solutions plus water was added primarily in the morning or early afternoon, depending on rates of evapotranspiration. The moisture level of each soil cup system was brought to a constant level each time solution and/or water was added. Seedlings in the soil cup system were supplied with nutrient solutions plus water every other day (controls and treatments). On alternate days control seedlings were supplied with water and treated seedlings were supplied with individual phenolic acid solutions, solutions containing mixtures of phenolic acids, or solutions containing phenolic acids and other organic compounds plus water. Initial solution pH values were adjusted between 5 and 5.8, unless otherwise specified. The reasons for choosing phenolic acids as the allelopathic agents are provided in Section 2.2.6. The number of phenolic acid treatments varied with experiment. For the soil columns, i.e., continuous-flow systems, seedlings were supplied with water or nutrient solutions (controls) or water or nutrient solution containing phenolic acids (treatments) by drip irrigation (Fig. 2.2; Blum et al. 1999a).

Finally, in a number of instances bioassays were carried out in a greenhouse (Hall et al. 1982, 1983) or in growth chambers (PPFD of 240–650 μmol/m^2/s; Waters and

Blum 1987; Blum and Rebbeck 1989; Klein and Blum 1990b; Holappa and Blum 1991; Lehman and Blum 1997, 1999a, b; Shafer et al. 1998) where materials and methods described above were modified to meet specific experimental objectives. For details see references cited.

2.2.2 Bioassay Species

Cucumber (*Cucumis sativus* cv. "Early Green Cluster") seedlings (Wyatt Quarles Seed Company, Raleigh, North Carolina) were selected as the bioassay species of choice after germination, radicle/hypocotyl growth, and/or seedling nutrient culture trials of 11 crop species (Blum et al. 1984; Blum and Dalton 1985). The other 10 species screened were not chosen because of their inconsistent germination rates, slow or variable growth rates under the light banks, inappropriate size or growth habit, inconsistent leaf characteristics, and/or low sensitivity to phenolic acids. In addition, over the years the following bioassay species were also used for seedling growth bioassays: bush snap bean [*Phaseolus vulgaris* L. "BBL-290" and "Oregon 91"], corn [*Zea mays* L. "Pioneer Hybrid 3369a"], morningglory [*Ipomoea hederacea* L.], red-root pigweed [*Amaranthus retroflexus* L.], and tomato [*Lycopersicon esculentum* Miller "wild type" and "*flacca*"] (Hall et al. 1982, 1983; Waters and Blum 1987; Holappa and Blum 1991; Bergmark et al. 1992; Blum et al. 1993; Pue et al. 1995; Lehman and Blum 1997; Shafer et al. 1998; Staman et al. 2001).

2.2.3 Soil Substrates

A_p- and B_t-horizon Cecil (Typic Hapludults, clayey, kaolinitic, thermic), Portsmouth (Typic Umbraquualts, fine loamy, mixed, thermic), and White Store (Vertic Hapludalfs, fine, mixed, thermic) soil materials were collected in the Piedmont and Coastal Plain of North Carolina, air-dried, sieved, and stored at room temperature in the laboratory. Soil samples were analyzed for minerals in the clay fraction, surface area of the clays, percent sand, silt and clay, cation exchange capacity, soil organic matter, soil pH, nutrient content, bulk density, and/or soil microbial populations (Table 2.1; Blum et al. 1987; Dalton et al. 1983, 1987, 1989a, b; Blum and Shafer 1988; Shafer and Blum 1991). Substrates were also tested for their ability to adequately grow cucumber seedlings. Because of the nature of the soil it was necessary to add sand to these substrates (1 unit of soil to 2 units of sand by weight) for good seedling growth, to minimize compaction, increase aeration, and water/solution percolation and distribution. White Store was eventually eliminated from consideration because of poor cucumber seedling growth. For best growth of cucumber seedlings nutrient additions were required. To maximize growth of cucumber seedlings nutrients were added either directly to the soil at the beginning of an experiment and/or added by the addition of Hoagland's nutrient solution during an experiment (Table 2.2; Hoagland and Arnon 1950; Blum et al. 1987, 1999a;

Table 2.1 Soil characteristics of Cecil, Portsmouth, and White Store soils[a]

Soil type	Cecil[b]		Portsmouth[b]		White Store[b]	
Horizon	A	B	A	B	A	B
Particle size (%)						
Sand	66.00	17.00	61.00	45.00	45.00	12.00
Silt	29.00	23.00	28.00	36.00	47.00	38.00
Clay	5.00	60.00	11.00	20.00	9.00	49.00
Organic matter (%)	3.70	0.20	3.90	0.07	2.40	0.34
Mineral constituents (% of clay fraction)[c]						
Kaolinite	75.00	75.00	39.00	27.00	49.00	20.00
Gibbsite	5.00	6.00	–	–	–	–
Fe_2O_3[d]	10.00	14.00	2.00	2.00	10.00	5.00
Mica	5.00	< 5.00	5.00	5.00	12.00	5.00
Quartz	–	–	1.00	10.00	12.00	10.00
Vermiculite[e]	5.00	< 5.00	–	–	–	–
Smectite/vermiculite[e]	–	–	45.00	55.00	17.00	60.00
Mineral constituents (% of total soil)[c]						
Kaolinite	3.75	45.00	4.29	5.40	3.92	10.00
Gibbsite	0.25	3.60	–	–	–	–
Fe_2O_3[d]	0.50	8.40	0.22	0.42	0.80	2.50
Mica	0.25	< 3.00	0.55	1.00	0.96	2.50
Quartz	–	–	1.10	0.20	0.96	5.00
Vermiculite[e]	0.25	< 3.00	–	–	–	–
Smectite/vermiculite[e]	–	–	4.95	11.00	1.36	30.00
Surface area (m^2/g)						
For clays	42.00	48.00	69.00	69.00	85.00	67.00
For total soil	4.00	33.70	2.40	8.20	4.00	35.80
CEC (millieq/g)	0.13	0.08	0.34	0.33	0.26	0.52
pH	6.20	5.40	6.60	4.85	5.20	4.65

[a]Sources of data: Dalton et al. (1987, 1989b), data used with permission of Soil Science Society of America; Dalton et al. (1983, 1989a).
[b]Cecil (Typic Hapludults, clayey, kaolinitic, thermic); Portsmouth (Typic Umbraquualts, fine loamy, mixed, thermic); White Store (Vertic Hapludalfs, fine, mixed, thermic).
[c]Estimates based on x-ray diffraction and/or differential thermal analysis except for Fe_2O_3.
[d]Total dithionite-reductable Fe expressed as Fe_2O_3 mainly goethite and hematite.
[e]Vermiculite = vermiculite/hydroxy-interlayered vermiculite, Smectite = smectite/hydroxy-interlayered smectite.

Klein and Blum 1990b; Shafer and Blum 1991; Pue et al. 1995). At various times pH of the soil was modified by addition of sodium hydroxide (Dalton et al. 1983), hydrochloric acid (Dalton et al. 1983), aluminum sulfate (Blum et al. 1987, 1989), or calcium hydroxide (Blum and Shafer 1988; Blum et al. 1989; Klein and Blum 1990b; Gerig and Blum 1991). When needed, soil substrates were also sterilized by several methods (Dalton et al. 1989a), but primarily by autoclaving (Dalton et al. 1983, 1989b; Blum and Shafer 1988; Blum et al. 1994, 1999a; Blum 1997, 1998; Staman et al. 2001). Finally, washed river sand, with a 2.5 cm surface layer of gravel

Table 2.2 Hoagland's nutrient solution[a]

Nutrient solution

To 750 ml of deionized water add the following ml of stock solutions (mix thoroughly after each addition)

Stock solution	ml
0.5 M KH_2PO_4	2
0.5 M KNO_3	10
0.5 M $Ca(NO_3)_2 \cdot 4H_2O$	10
0.5 M $MgSO_4 \cdot 7H_2O$	4
Supplementary solution	1
Chelated iron solution	1

Adjust pH to 5.8 or desired pH with NaOH and bring the final volume to 1 l with deionized water

Supplementary solution[b]

To 750 ml water add the following g of chemicals (make sure that each chemical is dissolved before adding the next one)

Substance	g
H_2BO_3	2.86
$MnCl_2 \cdot 4H_2O$	1.81
$ZnSO_4 \cdot 7H_2O$	0.22
$CuSO_4 \cdot 5H_2O$	0.08
H_2MoO_4	0.02

Bring final volume to 1 l with deionized water

Chelated iron solution[b]

Add 26.1 g of Na_2EDTA into 268 ml of 1 N KOH

Be sure that the EDTA is fully dissolved before adding 24.9 g $FeSO_4 \cdot 7H_2O$

Bring solution up to 1 l with deionized water

Bubble air through solution over night

Resulting solution should be clear and dark brown

[a]Source of data: Hoagland and Arnon (1950), data used with permission of ANR Communication Service and Information Technology, University of California.

[b]Stock solutions, supplementary solution and the chelated iron solution may be stored in a refrigerator for extended shelf life.

or perlite to minimize sand surface disturbance when water or solutions were added and to minimize sand surface algal growth, was also used as substrate (Waters and Blum 1987; Shafer et al. 1998).

2.2.4 Seedling Containers

Given the limited space under our light banks and the need to have adequate number of replicates per treatment, a number of containers of various shapes and sizes were tested to determine their adequacy for cucumber seedling growth over a 2-week period. For nutrient culture bioassays, 120 ml snap-cap Wheaton glass bottles were found to be an excellent choice (Fig. 2.3). Bottles were wrapped with aluminum foil and/or placed in wooden racks to exclude light from reaching roots and nutrient solutions. Seedlings were held in place by foam collars in a hole made in the snap

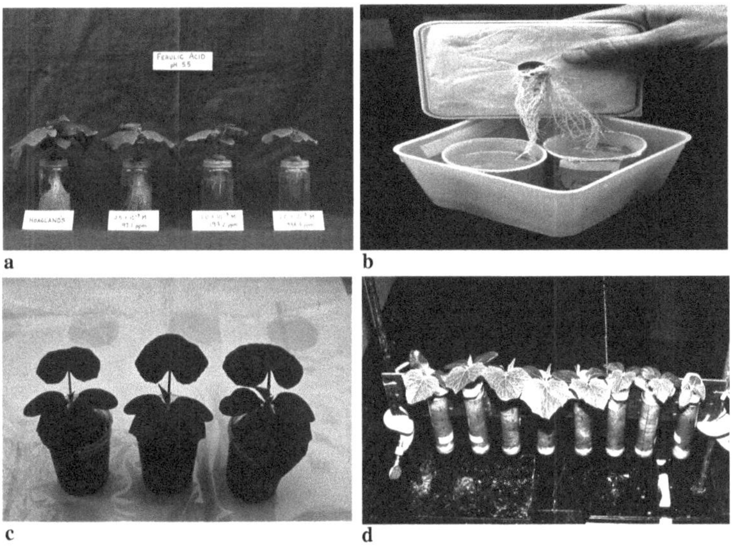

Fig. 2.3 Containers: **a** Wheaton glass bottles, **b** split-root systems, **c** soil cups, and **d** soil columns

cap (Blum and Dalton 1985). Wheaton glass bottles were reused. Before each experiment bottles were washed and autoclaved. For the split-root systems the seedlings were also suspended by foam collars, but in this instance through the lids of 700-ml plastic containers (Klein and Blum 1990a; Lyu and Blum 1990; Lehman et al. 1994; Fig. 2.3). The seedling root system for each 700-ml container was suspended into two smaller plastic containers (single container for some controls). One of the small containers was filled with 110 ml nutrient solutions and the other container with 110 ml of nutrient solution plus phenolic acid(s). Containers were covered with aluminum foil to exclude light. Small containers were thoroughly washed before each experiment. Unbreakable translucent 155 ml high-density polyethylene containers (cups) recommended for specimen collection (Fisher Scientific, US) proved to be ideal for the soil cup bioassays (Fig. 2.3; Blum et al. 1987). Cups without drainage holes were of sufficient size to hold 150 g of a given soil-sand mixture. Cups were placed into wooden racks to exclude light from reaching the sides of the containers. New cups were used for each experiment. For the continuous-flow systems soil columns composed of sterile 50-ml syringes containing a layer of cotton and 50 or 60 cm^3 of a soil-sand mixture were ideal (Fig. 2.3). New syringes were used for each experiment. Soil columns were suspended under the light banks so that solutions could be collected from the base of the columns (Blum et al. 1999a). Columns and collection vessels were protected from light with aluminum foil. However, because of experimental objectives, other types of containers (e.g., Styrofoam cups, quart jars, beakers, plastic containers, and plastic pots with liners) were also used (Waters and Blum 1987; Klein and Blum 1990b; Blum and Rebbeck 1989; Holappa and Blum 1991; Bergmark et al. 1992; Booker et al. 1992; Shafer et al. 1998; Lehman and Blum 1999a, b).

2.2.5 Sorption and Microbial Utilization Studies

Air-dried soil samples were passed through a 0.25-mm sieve. Materials not passing through the sieve without grinding were excluded. In some instances, the pH of the soil material was adjusted with NaOH or HCl and/or saturated with cations, e.g., Al^{3+} or Ca^{2+} (Dalton et al. 1983). One-tenth to one gram of soil material was placed into 15 ml Corex centrifuge tubes, capped with BACTI-CAPALL (Sherwood Medical Industries) polypropylene stoppers with flexible fins (fins permit gas exchange with minimum evaporation), and sterilized by several techniques (Dalton et al. 1989a), but predominately by autoclaving a total of two times (3 days apart, 121°C at 1.2 kg/cm^2 for 15 min). Various amounts of solutions (e.g., water, nutrients, and phenolic acids) were added aseptically to the tubes, tubes were vortexed, and were then stored in the dark for various time periods (hours to weeks) before soils were extracted to determine amounts of phenolic acids left in solution, and reversibly and irreversibly sorbed to the soil materials (Dalton et al. 1983, 1987, 1989a, b; Blum et al. 1994; Blum 1997). In some instances soils were inoculated with microbes by the addition of soil-water extracts to determine microbial utilization of phenolic acids (Blum 1997, 1998). Soil extracts to inoculate soil materials were obtained by treating 150 g air-dried soil (soil:sand mixture, 1:2 by weight) with 5 ml of 0.5 mM phenolic acid(s), 10 ml of full strength Hoagland's solution, and 15 ml water. After 4 or 5 days in the dark at 32°C, the soil sand mixture was mixed with 300 ml water in a beaker. Soil particles were allowed to settle in the beaker before the resulting solution was filtered through Whatman No. 1 filter paper in a Buchner funnel. The resulting filtered extract was used to inoculate soil samples. Controls were supplied with an equivalent amount of filter sterilized (0.2 μm membrane filter) soil extract.

2.2.6 Phenolic Acids

Historically, simple phenolic acids have been the most frequently identified allelopathic agents (see literature reviews by Rice 1974, 1979, 1983, 1984, 1986). One would assume this was partly because of the fact that the necessary technology to isolate, identify, and quantify phenolic acids, even though crude in the early days, was readily available to most researchers. Furthermore, simple phenolic acids, such as the benzoic acid and cinnamic acid derivatives serve a variety of plant and ecosystem functions and are widespread in higher plants (Fig. 2.4; Bates-Smith 1956; Harborne 1982, 1990; Goodwin and Mercer 1983; Siqueira et al. 1991). The ubiquitous distribution in nature and their apparent rapid turnover rates in soils, however, have lead to some controversy as to the importance of phenolic acids in plant–plant allelopathic interactions (Schmidt 1988; Schmidt and Ley 1999; Blum 2004, 2006). Finally, the behavior of phenolic acids in soil systems are somewhat similar to the behavior of a whole host of other organic acids (e.g., acetic acid, butyric acid, citric acid, formic acid, fumaric acid, lactic acid, malonic acid, tannic acids and tartaric

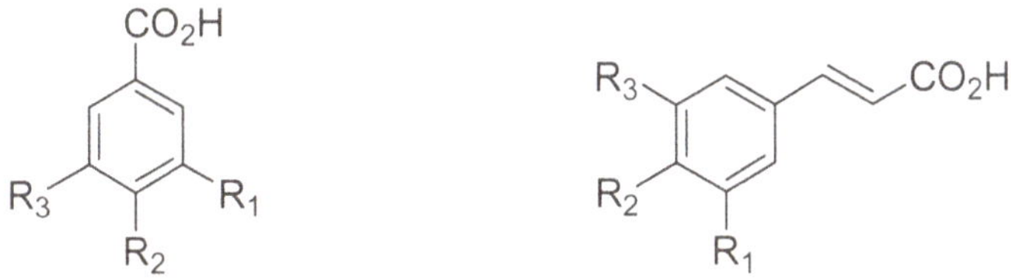

R1	R2	R3	
H	OH	H	*p*-hydroxybenzoic acid
OH	OH	H	protocatechuic acid
OMe	OH	H	vanillic acid
OMe	OH	OMe	syringic acid

R1	R2	R3	
H	OH	H	*p*-coumaric acid
OH	OH	H	caffeic acid
OMe	OH	H	ferulic acid
OMe	OH	OMe	sinapic acid

Fig. 2.4 Some common simple plant phenolic acids, cinnamic acid derivatives on the right and benzoic acid derivatives on the left, where H equals hydrogen, OH equals hydroxy, and OMe equals methoxy

acid) that have also been identified as potential allelopathic agents (Patrick 1971; also see Rice 1984). To my mind all these aspects plus their availability in pure form, water solubility, reliable extraction procedures for plant and soil materials, and isolation and identification by High Performance Liquid Chromatographic (HPLC) analysis made phenolic acids intriguing and ideal candidates for in-depth study.

2.2.7 Phenolic Acid Solutions

Solubility of phenolic acids in water increases with solution temperature and pH. At 25°C and a pH of 5.0 roughly 15 mM stock solutions can be achieved for the following cinnamic acid derivatives (Fig. 2.4), caffeic acid (3,4 dihydroxycinnamic acid, 180.2 mol wt), ferulic acid (4-hydroxy-3-methoxycinnamic acid, 194.2 mol wt), *p*-coumaric acid (*p*-hydroxycinnamic acid, 164.2 mol wt), and sinapic acid (4-hydroxy-3,5-dimethoxycinnmaic acid, 224.21 mol wt) and benzoic acid derivatives, *p*-hydroxybenzoic acid (138.1 mol wt), protocatechuic acid (3,4-dihydroxybenzoic acid, 154.1 mol wt), syringic acid (4-hydroxy-3,5-dimethoxybenzoic acid, 198.2 mol wt), and vanillic acid (4-hydroxy-3-methoxybenzoic acid, 168.2 mol wt). Phenolic acids, except for sinapic acid, were purchased from Sigma Chemical Company, St. Louis Missouri. Sinapic acid was obtained from Aldrich Chemical Company, Milwaukee, Wisconsin. Phenolic acid stock solutions were made by placing the appropriate amount of phenolic acid or acids into a volumetric flask with a stir bar and distilled water. Solutions should be made in semi-darkness (Katase 1981a). Drops of 0.5–1 M sodium hydroxide were added slowly to the volumetric flask to bring the pH of the solution to approximately pH 4.5 while its contents were rapidly stirred and brought close to a boil. Concentrations and the number of drops required to bring solution to a pH of 4.5 can readily be determined in a preliminary study. Once phenolic acids were completely dissolved, the solution was allowed to come to room temperature in the dark before final pH and volume were adjusted with 0.1–1 M NaOH. For nutrient culture, sand culture, or continuous flow studies phenolic acid stock solutions were combined with the appropriate strength of nutrient solution to generate the desired phenolic acid concentrations and nutrient strengths. For soil cup studies phenolic acid stock solutions were diluted with water to generate the desired phenolic acid concentration in the soil (μmol/g soil). The volume of solution added to the soil systems was constant for any given experiment. Solutions were filter sterilized (0.2 μm membrane filter) when and where needed to eliminate microbial contamination.

2.2.8 Solution Additions to Seedling Systems

Nutrient and water requirements for consistent growth and development under the light banks were determined for cucumber seedlings in nutrient culture (Blum and Dalton 1985), in soil cultures (Blum et al. 1987; Klein and Blum 1990b; Pue et al. 1995), and in soil column culture (Blum et al. 1999a). For the nutrient culture bioassays using Wheaton glass bottles, 110 ml of single strength Hoagland's nutrient solution (Hoagland and Arnon 1950; Table 2.2) was adequate for cucumber seedling growth over a 2-week period (Blum and Dalton 1985). Water lost through transpiration was added back each day. For experiments with multiple phenolic acid treatments the entire nutrient solution volume plus or minus phenolic acid or acids was exchanged every other day. The pH of all solutions was adjusted to a range from pH 4 to pH 7, depending on experiment, with sodium hydroxide. For some experiments MES buffer was used to stabilize the pH of the nutrient/phenolic acid solutions (Blum et al. 1984, 1985a, b; Shann and Blum 1987a; Booker et al. 1992; Lehman and Blum 1999b). Phenolic acid solutions before pH adjustments are very acidic (< 3.6). A pH range of 6.0–6.5 is recommended for best growth of cucumber seedlings (Hughes et al. 1983) but effects of phenolic acids require an acidic pH range (Blum et al. 1985b), thus the use of pH 5.0–5.8 for the majority of experiments, was a compromise. However, it should be noted that even in the presence of a buffer such as MES, which only slows the rate of change, pH values of the Hoagland's solutions move towards the neutral end over time because of the action of seedling roots and microbes (Blum et al. 1985b), the older the seedling the faster the change in pH. For good growth nutrient solutions for growth chamber bioassays had to be reformulated (Blum and Rebbeck 1989; Holappa and Blum 1991). For some nutrient uptake studies treatment solutions were also reformulated (Lyu and Blum 1990; Lyu et al. 1990; Bergmark et al. 1992; Booker et al. 1992; Lehman and Blum 1999a, b). For split-root systems water was also added each day as needed to the two containers or complete solution changes were made when portions of the root system were treated with phenolic acid(s) (Klein and Blum 1990a; Lyu and Blum 1990; Lehman et al. 1994). In some instances solutions were aerated (Lyu and Blum 1990; Lyu et al. 1990; Holappa and Blum 1991; Bergmark et al. 1992; Booker et al. 1992; Lehman and Blum 1999a; Blum and Gerig 2005). Because soils contain complex mixtures of phenolic acids, both seeds and seedlings were not only treated with individual phenolic acids to determine their individual inhibitory activity, but also with mixtures of phenolic acids (Blum et al. 1984, 1985a; Lyu et al. 1990; Lehman et al. 1994; Blum 1996).

Complete solution changes for multiple phenolic acid treatments were used because phenolic acids supplied to seedlings in the nutrient culture system disappeared from the nutrient solution within 24–48 h (Blum and Dalton 1985; Blum and Gerig 2005). This was due to microbial metabolism, physical breakdown, and/or root uptake. Since we did not want to confound nutrient and phenolic acid effects, complete solution changes were made. An additional benefit of this approach was to reset phenolic acid concentrations to the initial treatment levels for each solution change. This was important since recovery of seedling processes occurred rapidly after phenolic acid depletion (Blum and Dalton 1985; Blum and Rebbeck 1989; Blum and Gerig 2005).

For the soil bioassays the amount of double strength Hoagland's nutrient solution added varied with experiment but generally ranged from 6 to 8 ml every other day. To balance nutrition and/or to further reduce the volume of the nutrient solution added to the soil systems, nutrients were mixed directly into the soil (Blum et al. 1987; Blum and Shafer 1988) and/or up to quadruple strength Hoagland's nutrient solution was added every other day (Klein and Blum 1990b; Pue et al. 1995). Phenolic acid solutions and nutrient solutions were given on alternate days. One to 10 ml, depending on experiment, of water (control) and/or phenolic acid solution (treatment) was supplied to the soil systems. Because soils contain complex mixtures of phenolic acids and other organic compounds, both seeds and seedlings were not only treated with single phenolic acids to determine their individual inhibitory activity, but also with mixtures of phenolic acids (Blum et al. 1989; Gerig and Blum 1991; Blum 1996) and/or mixtures of phenolic acids and other organic compounds (Blum et al. 1993; Pue et al. 1995). Amounts of phenolic acids and other water soluble organic compounds added to soil systems were in a constant volume for any given experiment (Water insoluble organic compounds were mixed directly into the soil). The pH values of all solutions added to soil systems were adjusted to a range of pH 5.0–pH 7.5, depending on experiment, with sodium hydroxide. After addition of phenolic acid(s), nutrient solutions, nitrogen solutions, and/or solutions containing other organic compounds, water was added to bring the cup systems to an initial constant weight each day.

Maintaining a specific range of soil water for the seedling-soil-systems was essentially impossible given the size of the cup system and the variation in seedling size and evapotranspiration. We, therefore, chose to supply water once a day at the same time to all systems. Bringing water levels to less than 20 g water/150 g soil each day required that water be added more than once during the day to prevent wilting of cucumber seedlings. For example, in Cecil A soil-sand systems, the first sign of wilting by cucumber leaves occurred whenever soil water in the system reached 7.24 ± 0.17 g water/150 g soil (Blum and Gerig 2006). Bringing water levels to more than 25 or 30 g water/150 g soil each day resulted in standing soil surface water for several hours during each 24-h watering cycle, the time varying with seedling age, and with phenolic acid treatment. The level and time of soil surface standing water also varied with soil type. The amount of water added daily actually declined slightly over time as seedling weight increased. It also varied some with inhibition of seedlings by phenolic acids. Recall that water added at the beginning of each 24-h water cycle was based on seedling-soil-system weight. However, this error was relatively small. For example, in the Cecil A soil-sand system this amounted to less than 1.5 g at the termination of experiments and less than 0.5 g between controls and *p*-coumaric acid treatments (Blum and Gerig 2006).

Multiple additions of phenolic acids were used because phenolic acid concentrations in soil decline rapidly after each addition of phenolic acids (Blum et al. 1987; Blum and Gerig 2006). This was due to microbial metabolism, physical breakdown, root uptake, and/or soil particle sorption. Recovery of seedling processes, although considerably slower than in nutrient culture, also occurred in seedling-soil systems (Blum et al. 1987; Blum and Gerig 2006). To maintain inhibition for extended time periods multiple additions of phenolic acids were required.

For the soil columns (Fig. 2.2), water, nutrient solution, or various concentrations of phenolic acid in water or nutrient solution was supplied to the top of each column at 2–3.5 ml/h with a cassette pump (Blum et al. 1999a). Solutions were collected from the bottom of the column for 30 min at 12-h intervals during phenolic acid treatment to estimate the amount of phenolic acid lost (leached) from the systems.

For sand culture, which was carried out in growth chambers, phenolic acids were combined with Phytotron nutrient solution (Downs and Thomas 1983) at the desired concentrations and supplied once or twice a day for various time periods (Waters and Blum 1987; Shafer et al. 1998). Styrofoam cups (240 ml) or plastic pots (15.2 cm diameter, 1,650 ml) with screens and holes at the bottom were used (i.e., open systems). For one study enough phenolic acid-nutrient solution was added twice a day to completely displace the void volume of the sand in the container (Shafer et al. 1998). For the other study treatments were given at various times during the plants life cycle (Waters and Blum 1987). Four hundred ml of phenolic acid-nutrient solution was added in the morning to each pot of the first day and a plastic liner around the pot was used to capture the solution coming from the pot and the captured solution plus sufficient nutrient solution to bring the volume back to 400 ml was recycled in the afternoon. On the second day the pots were flushed with deionized water in the morning and watered with nutrient solution in the morning and the afternoon. This sequence was repeated on the third and fourth day. The plastic liner was removed on the fourth day. Thereafter the pots were supplied with nutrient solution twice a day.

2.2.9 Phenolic Acid Extraction Procedures

Simple phenolic acids are colorless solids when pure, but usually oxidize and become dark on exposure to air (Robinson 1967). Water solubility increases as the number of hydroxyl side groups, solution temperature and/or solution pH values increase. Solubility is generally very high in polar organic solvents. However, since under basic conditions their rate of oxidation increases, prolonged treatment with strong alkali should be avoided.

A range of extractants and extraction procedures has been used to extract phenolic acids from soil (Dalton 1999). Many of these extractants and extraction procedures, however, recover phenolic acids that are not directly involved in plant–plant allelopathic interactions (e.g., phenolic acids sorbed in the recalcitrant organic matter). Thus considerable efforts were made to identify extraction procedures that would provide reasonable estimates of available phenolic acids ("free" phenolic acids in soil solutions and reversibly sorbed phenolic acids on soil particles) in soils (Dalton et al. 1983, 1987, 1989a, b; Blum et al. 1994; Blum 1997; Dalton 1999).

We settled primarily on water and 0.25–0.5 M neutral EDTA (ethylenediaminetetraacetic acid; chelating agent, pH 7) extractions (for justifications see Section 2.4.3; Blum et al. 1994) and a water-autoclave procedure (for justifications see Section 3.4.1; Blum et al. 1991, 1992). Among others used were methanol and sodium hydroxide (Dalton et al. 1987), sodium acetate (Blum et al. 1987), Mehlich III (Dalton et al. 1989a, b), and citrate plus or minus imidazole, KCL, or dibasic sodium phosphate (Blum 1997) extractions. Since the details regarding each specific extraction procedure are complex in that soil samples extracted ranged from less than 1 g to more than 30 g, concentration of extractants ranged from μmol to pure solvent, pH of the extractants ranged from 4.5 to 12, temperatures of extraction procedures ranged from room temperature to 121°C, and pressure of extraction procedures ranged from ambient to 1.2 kg/cm^2, exact details are not provided here. However, if such details are of interest to the reader he or she should look at the references cited for details. What should be remembered, however, is that for water based extractions the pH of the soil-extractant slurry (for our soils neutral being the best) and the time of contact between soil and extractant (for our soils 2.5–5 h appeared to be adequate) are crucial (Dalton et al. 1987; Blum et al. 1994, 1997). Thus sufficient strength and quantity of an extractant should be used to overcome the buffering capacity of the soil sample and adequate time of contact between soil and extractant should be used to fully recover the available phenolic acids in soil.

For our soils water, EDTA, citrate, NaOH, etc. extractions were carried out on individual samples or subsamples of soils with or without amendment phenolic acid(s). Thus the recovery for each extractant comprised the sum of the "free", reversibly sorbed, and/or fixed (not immediately available to roots or microbes) phenolic acids that were recoverable by a given extractant. Differences between water extractions (primarily "free" phenolic acids) and EDTA or citrate ("free" and reversibly sorbed phenolic acids) extractions were utilized to estimate reversibly sorbed phenolic acids. Differences between EDTA or citrate extractions and NaOH extractions were utilized to estimate the fraction of fixed phenolic acids that could be recovered by NaOH. For additional details see Section 2.4.3. This approach has been criticized because "free" phenolic acids were not removed from soil samples before reversibly sorbed phenolic acids were extracted (Ohno and First 1998). However, in a preliminary study extracting our soils by the traditional method (removing "free" phenolic acids before extracting sorbed phenolic acids) and our method (using different soil samples for each extractant without first removing "free" phenolic acids) made little difference in the resulting data (Blum 1998).

Citrate extractions of soils deserve some additional comments. Citrate extractions have similar individual phenolic acid recovery rates as neutral EDTA extractions from soil, at least for Cecil soils and utilizing HPLC analysis (Blum 1997). However, citrate, unlike EDTA, does not reduce the Folin & Ciocalteu's phenol reagent used to estimate total available phenolic acid content of soil extracts (Blum et al. 1992; Blum 1997). On the negative side, citrate extractions of soils with high organic matter can lead to unreliable estimates of total available phenolic acids (Ohno and First 1998). In fact, Ohno and First suggest that the large quantities of organic matter extracted by citrate mask absorbance due to any extracted simple phenolic acids and thus the Folin & Ciocalteu's phenol reagent method should not be used on citrate-soil extracts.

To extract phenolic acids in plant tissues/residues ground plant tissues/residues were extracted with water, EDTA, citrate plus or minus imidazole, KCl, or dibasic sodium phosphate and the water-autoclave procedure (Blum et al. 1992; Blum 1997). For procedures used for ^{14}C-labeled phenolic acids see Section 2.2.10

2.2.10 Quantification of Individual Phenolic Acids

Nutrient culture samples were filter sterilized (0.2 μm membrane filter) and phenolic acids in the nutrient solution were isolated and quantified with a Waters (Milford, Massachusetts) fully automated HPLC equipped with a model 484 absorbance detector set at 254 nm and a 810 Chromatographic Workstation (Dynamics Solutions, Ventura, California). A Waters 5-μm-particle size Nova-pak C_{18} Radial Pak cartridge in a RCM-100 cartridge holder was used to separate/isolate the phenolic acids of interest. Nutrient culture samples were analyzed using a single (isocratic) combination of two mobile phases to separate phenolic acids: (A) 2% methanol, 0.25% ethyl acetate, 0.5% acetic acid, and 97.25% water and (B) 80% methanol, 1% ethyl acetate, 2% acetic acid, and 17% water. Combination of A and B were adjusted for best separation of phenolic acids of interest. Depending on the complexity of phenolic acids in a set of samples, runs of 15–30 min were used. The flow rate of the mobile phase was 0.5 ml/min. Plant tissues, plant residues, and soil extracts were handled in a similar manner to nutrient culture samples except that the two mobile phases varied over time. Gradients starting with 92% (A) and ending with 66% (A) were used over the first 40 min of a 60-min run. Identification and quantification were confirmed by comparing retention times and areas with those of the appropriate standards (prepared in deionized water) and by spiking unknown samples with standards. Standard phenolic acids were obtained from Sigma Chemical Company (St. Louis, Missouri) or Aldrich Chemical Company (Milwaukee, Wisconsin). Additional confirmation of the identity of extracted phenolic acids was obtained with paper or thin layer chromatography and by UV spectral comparisons in ethanol and ethanol plus sodium hydroxide (Harborne 1984).

To extract and determine seedling distribution of ^{14}C-labeled ferulic acid or *p*-hydroxybenzoic acid taken up by excised roots or seedlings freeze-dried and ground root, stem, cotyledon or leaf tissues (approximately 50 mg) were treated with 0.5 ml peroxide (30%) in scintillation vials and then allowed to bleach and dry under a light bank. Once dry, ethanol (2 ml, 95%) was added to the vials which were then capped and incubated at 40–50°C overnight before radioactivity was determined by liquid scintillation spectrophotometry (Shann and Blum 1987b). To determine how much of the radioactivity in the tissues represented ferulic acid or *p*-hydroxybenzoic acid, tissues were extracted with water or methanol. Ferulic acid or *p*-hydroxybenzoic acid in the extracts was isolated with a High Performance Liquid Chromatograph (HPLC), peaks were collected, and their radioactivity was determined by liquid scintillation spectrophotometry (Shann and Blum 1987a).

2.2.11 Rhizosphere and Soil Microbial Populations

Microbial populations in the rhizosphere were extracted as follows: root systems of each plant were dipped briefly into 250 ml of sterile deionized water to dislodge large aggregates of soil material. The entire root system with adhering soil material was then transferred to 99 ml blank sterile diluent (0.1% aqueous agar) containing 1 g of sterile sand and shaken for 30 min on a wrist-action shaker (Shafer and Blum 1991). Subsamples of rhizosphere extracts were taken and roots were then removed, rinsed, oven dried, and weighed. Microbial populations in bulk soil were extracted as follows: soil materials were diluted 1:1 (soil material: 0.1% sterile aqueous agar), and shaken on a wrist-action shaker for 15 min (Blum and Shafer 1988). Subsamples of rhizosphere and bulk-soil extracts were subsequently diluted (tenfold series) to 10^{-8} or 10^{-10} with 0.1% sterile aqueous agar. The resulting suspensions were vortexed and assayed by the plate-dilution frequency technique (Harris and Sommers 1968).

Petri plates with agar media for specific types of microorganisms were inoculated with 8 individual droplets of 0.01 ml for each suspension (up to 2 suspensions per petri dish) and held for 6 days at 30°C in the dark before microorganisms were enumerated (number of positive droplets). The colony-forming units (CFU)/g root dry weight or soil dry weight were determined from tables provided by Harris and Sommers (1968). The smallest population of any microbial group that could be enumerated was 20 CFU/g of root or soil.

Medium for enumeration of actinomycetes (ACT) was basal medium that consisted of (all ingredients in g per liter of deionized water): KH_2PO_4(1.00), $MgSO_4 \cdot 7H_2O$ (0.40), $CaCl_2 \cdot 2H_2O$ (0.13), NaCl (0.10), $FeCl_3$ (0.01), KNO_3 (0.50), Delvocid (50% natamycin, an antimycotic; GB Fermentation Industries, Inc. Charlotte, NC; 0.05) and agar (15). Medium for enumeration of bacteria consisted of ACT amended with glucose (1.00). Medium for enumeration of fungi consisted of ACT without Delvocid but amended with glucose (1.00), rose bengal (0.05) and chloramphenicol (0.10). Medium to enumerate bacteria that could utilize phenolic acids as a sole carbon source consisted of ACT medium containing 0.5 mM phenolic acid or mixtures of phenolic acids as the sole carbon source (Blum and Shafer 1988; Shafer and Blum 1991; Blum et al. 2000). All media were adjusted with NaOH to pH 7.0 before autoclaving. Media did not change more than 0.1 pH units during sterilization process. Microbial populations were expressed as the logarithm ($\log_{10}$) of the number of colony-forming units per gram of dry root or soil. Conclusions concerning statistical significance of treatment effects were based on analyses conducted on log-transformed data.

2.2.12 Measurements

Given the dynamic nature of plants, or for that matter all organisms, the question as to what to measure and when those measurements should be taken is an important one. Plant biochemists, molecular biologists, and physiologists tend to look for primary sites of action (e.g., the initial site or sites that is or are affected by an environmental variable of interest that will subsequent lead to a cascade of other effects) and ecologists, agronomists, and weed scientists tend to look at one of the resulting cascade effects. To identify primary sites of action requires that responses be detected and measured very rapidly after initial contact, ideally let's say on the order of seconds or minutes. The focus would thus be on membranes (e.g., membrane potentials or structure) or cellular processes (e.g., enzyme rates, respiration, and photosynthesis). Any significant delay after a primary effect has taken place, on the order of hours or days, will make it very difficult to identify primary sites of action, because immediately after a primary effect is experienced by any living organism there are secondary, tertiary, etc. effects (generally referred to as secondary effects). Thus changes in transpiration, nutrient uptake, leaf area, and root or shoot biomass, for example, represent secondary effects that are far removed from a primary effect. Although we made an attempt, where possible, to identify potential primary sites of action, that was not the focus of our research. After all plant–plant allelopathic interactions are generally observed on a scale of days, weeks, or months. Having said that, however, the shorter the time periods before measurements are taken after treatment the better are ones chances of characterizing the dynamic processes which ultimately lead to observable plant–plant allelopathic interactions. The down side of relying too much on short time intervals for such measurements is that plants have considerable "buffering capacity" and can readily acclimate and thus short term responses may or may not lead to observable long term effects. Note: Although primary effects may occur within seconds or minutes after contact they may not occur within seconds or minutes after treatment since time may be required for the treatment variable of interest to reach the primary site. This is one reason why plant biochemists, molecular biologists, and physiologists study effects at the molecular or cellular level.

Actual measurements can be made in a destructive manner (e.g., replicated systems can be destructively harvested at various time intervals) or in a non-destructive manner (e.g., measurements are taken on the same system or its components repeatedly over time). We used both approaches. For example, for our cucumber systems we monitored, radicle/hypocotyl length, transpiration, leaf area, and water depletion using non-destructive procedures and shoot, root and seedling dry weights, seedling nutrient and/or phenolic acid content, soil sorption, phenolic acid content in soil solution, and soil and rhizosphere microbial populations using destructive procedures. Partial destructive procedures were utilized for sampling nutrients and phenolic acids in nutrient solutions, e.g., small subsamples were taken at various time intervals. Wherever possible, non-destructive procedures were used. The pros for utilizing non-destructive procedures are:

a. The pattern of response and recovery after treatment can be more readily observed.
b. It takes fewer replicates and thus takes less space and resources for each experiment, and
c. The time involved in obtaining individual measurements is generally much less.

The cons for utilizing non-destructive procedures are:

a. The lower number of replicates may or may not be representative of the entire population under study, and
b. The data collected for the same unit (e.g., leaf area) measured repeatedly over time violate the assumptions of standard statistical analyses.

A number of measurements related to inhibition and recovery of seedling processes and changes in soil processes were made to determine how seedlings/seedling-soil systems responded to phenolic acid treatments. Among the processes/factors measured were: transpiration and water utilization (Blum et al. 1985a, b; Lyu and Blum 1990; Holappa and Blum 1991; Blum and Gerig 2005), water relations (Booker et al. 1992), abscisic acid levels (Holappa and Blum 1991), root growth (Blum and Rebbeck 1989), uptake and/or efflux of nutrients, such as ammonium, nitrate, potassium, and phosphorous (Lyu and Blum 1990; Lyu et al. 1990; Bergmark et al. 1992; Booker et al. 1992; Lehman and Blum 1999a, b), root uptake of phenolic acids (Blum et al. 1985a; Lehman and Blum 1999b; Blum and Gerig 2005), photosynthesis and carbon allocation to roots (Blum and Rebbeck 1989), lignin biosynthesis (Shann and Blum 1987b); leaf development (Blum and Dalton 1985), absolute and relative rates of leaf expansion (Blum and Dalton 1985; Blum and Gerig 2005), shoot and root biomass (Blum and Dalton 1985), root length and number (Blum and Rebbeck 1989; Klein and Blum 1990a), evapotranspiration and minimum soil water levels (Blum and Gerig 2006), soil respiration (Blum et al. 1987; Blum and Shafer 1988), bulk-soil and rhizosphere microbial populations (see Section 2.2.11; Blum and Shafer 1988; Shafer and Blum 1991), utilization of phenolic acids by soil microbes (Blum 1998, 2004, 2006), and available (i.e., "free" and reversibly sorbed) and irreversibly sorbed phenolic acids on/in soil and plant residues (see Sections 2.2.9 and 2.2.10; Blum 1997; Blum et al. 1994, 1999b). For precise details regarding how all these processes were measured please see previous method sections and/or references cited. However, a few general descriptions regarding some of the more common processes or factors measured are provided here.

Transpiration on individual leaves was determined with a Li-Cor Steady State Porometer (model Li-1600) or by monitoring water utilization (Blum et al. 1985a, b; Lyu and Blum 1990; Blum and Gerig 2005). Water utilization in nutrient culture was determined by measuring solution remaining in snap-cap bottles and subtracting that from the original volume or by converting the height of the water level at a given point in time in the snap-cap bottle to ml of water and then subtracting that value from the starting ml value (Blum et al. 1985a; Blum and Gerig 2005). One example of a model used to convert water level in a snap-cap bottle to ml of water was the following: ml in bottle $= -10.549 + (1.523 \times \text{solution level})$, $p < 0.001$, $r^2 = 0.99$, where solution level is in mm (Blum and Gerig 2005).

For seedling-soil-cup systems water utilization was expressed as evapotranspiration since evaporation from the soil and the transpiration of seedlings could not be separated. Evapotranspiration of seedling-soil-cup system was determined by the decline in system weight over time and assuming that a 1 g decline over a given period for a system equaled 1 ml of water evapotranspired (Blum and Gerig 2006).

Seedling uptake of phenolic acids for the nutrient culture system was determined by depletion from nutrient solution (Blum et al. 1985a; Shann and Blum 1987a; Lyu et al. 1990; Lehman and Blum 1999b; Blum and Gerig 2005). Small subsamples (2 ml or less) were taken at various time intervals and the resulting subsamples were analyzed by HPLC analysis (see Section 2.2.10 for details). Initial microbial breakdown products of phenolic acids were also monitored (Blum et al. 1985a; Blum and Gerig 2005). Uptake was also determined by the uptake of [U-ring-^{14}C] labeled phenolic acids by excised roots and seedlings (Shann and Blum 1987a; see Section 2.2.10). In soil systems depletion of phenolic acids (combinations of seedling uptake, sorption, and/or microbial utilization) and the initial phenolic acid products of microbial utilization/breakdown of phenolic acids were monitored by extracting soils and analyzing extracts by HPLC analysis at various time intervals after phenolic acid amendments (see Sections 2.2.9 and 2.2.10; Blum et al. 1987; Blum 1998, 2004, 2006).

Net uptake and/or efflux of nutrients, i.e., phosphorous, potassium, nitrate, ammonium, chloride, and sulfate (Lyu et al. 1990; Lyu and Blum 1990; Bergmark et al. 1992; Booker et al. 1992; Lehman and Blum 1999a, b) were also based on depletions or increases, respectively, for treatment solutions containing the nutrient(s) in question, with or without phenolic acid, and calcium to maintain membrane function and metabolic regulation. Phosphorous was determined colorimetrically (Taussky and Shorr 1953). Potassium was determined by atomic absorption spectrophotometry or by flame photometry. Nitrate and sulfate were determined with a Dionex 2010-I ion exchange chromatograph equipped with an AS4A anion exchange column with a micromembrane suppressor. Nitrate was also determined using a manual modification of the procedure of Lowe and Hamilton (1967). Ammonium was determined colorimetrically by a modified sodium hypochlorite assay (Cataldo et al. 1974). Chloride was determined with a chloride titrator.

Equations based on leaf or cotyledon length, width, and area were developed for estimating areas of true leaves (Blum and Dalton 1985; Waters and Blum 1987) and cotyledons (Blum and Gerig 2005) non-destructively from leaves and cotyledons that were destructively harvested. Leaf and cotyledon areas were determined with a Li-Cor model Li 3000 leaf area meter. Examples of equations for leaf and cotyledon areas of cucumber seedlings were as follows: leaf area in $cm^2 = -1.457 + 0.008 (L \times W)$, $p < 0.001$, $r^2 = 0.98$ where L (leaf length) and W (leaf width) were in mm, and cotyledon area in $cm^2 = 1.293 + 0.006 (L \times W)$, $p < 0.001$, $r^2 = 0.76$ where L (cotyledon length) and W (cotyledon width) were in mm. Utilizing such equations, leaf and/or cotyledon areas and the growth rates of individual leaves and/or cotyledons were estimated and monitored for cucumber seedlings in the presence and absence of phenolic acids (Blum and Dalton 1985; Blum and Gerig 2005). The absolute growth rates of leaf expansion (cm^2/seedling/unit time) for cotyledons (modified leaves) and/or true leaves were determined as follows: leaf area at $time_{x+1}$ – leaf area at $time_x$. The relative growth rates of leaf expansion (cm^2/cm^2/unit time) were determined as follows: ln (leaf area at $time_{x+1}$) – ln (leaf area at $time_x$). For a discussion on calculating growth rates, see Radford (1967). An additional benefit of utilizing leaf area of cucumber seedlings as a primary variable was that shoot and root dry weight, total and average root length, and root number were significantly related to leaf area (Blum and Rebbeck 1989).

The non-destructive monitoring of seedling leaf area requires some additional comments since each leaf grows only until it reaches its maximum size. In addition, growth rates tend to decline as a leaf reaches its maximum size. Effects are thus detectable only for expanding immature leaves. As a seedling ages the fully expanded non-responsive leaves make up a larger portion of the total seedling leaf area and thus reduce the magnitude of the effect observed for seedlings. Care must, therefore, be taken in interpreting and comparing the magnitude of observed effects for total seedling leaf areas over time. A partial way around this dilemma is to use absolute and relative rates of leaf expansion (i.e., growth rates) instead of total seedling leaf area. Growth rates emphasize the effects observed for younger expanding leaves and reduce the dampening effects of older non-expanding leaves.

2.2.13 Data Analyses

Data were analyzed using SAS systems (SAS Institute Inc. 1999) primarily employing analysis of variance and regression analysis. Statistical significance was based on alpha ≤ 0.05. To determine general patterns or trends in the data, linear and polynomial regressions were used. For such curve fitting no attempts were made to determine the exact curves or to account for the potential variability in the data.

Where appropriate, models were developed with the joint goals of providing a parsimonious explanation of the data while maintaining maximum predictive power. Initial model screening and computation of the r^2 statistic were handled using SAS/GLM. However, for many of the analyses the data contained observations on the same unit measured repeatedly over time. This created a violation of the assumptions for standard statistical techniques and the need to use methodologies for handling models with repeated measures. The tool used in this case was the SAS/MIXED procedure (Littell et al. 1996).

Some comments regarding the analysis of inhibitory effects of mixtures of compounds are warranted. In these instances the resulting models were nonlinear and required the use of the SAS/NLIN procedure. For inhibitory effects of mixtures of phenolic acids and mixtures of phenolic acids and other organic compounds a joint action model, modification model, or multiplicative model were utilized depending on types of compounds in the mixture (Gerig et al. 1989; Gerig and Blum 1993). For the joint action model the compounds in the mixture must show reciprocity. Relative potency of a compound relative to a standard compound was employed for this model. This model can be generalized to three or more compounds and is the only model that allows for the determination of synergistic, additive or antagonistic behavior. The modification model requires that a second compound that is not inhibitory changes the effective inhibitory concentration of the first compound. If the compounds involved are all inhibitory by themselves (i.e., independent in action) the multiplicative model is the most appropriate. However, when two compounds interact to form a new compound, then none of these models are appropriate.

2.3 Research Objectives

The primary research objectives were as follows:

a. To determine the effects and the duration of effects of phenolic acids on seedlings in nutrient culture.
b. To determine the effects of seedlings, mixtures of phenolic acids, and microbes on phenolic acid concentrations in nutrient culture.
c. To determine the interactions of phenolic acids with sterile and non-sterile soils.
d. To determine the effects of phenolic acids on bulk-soil and rhizosphere-microbial populations.
e. To determine the effects and the duration of effects of phenolic acids on seedlings in soil culture.
f. To determine relationships between phenolic acid-utilizing microbes and phenolic acid inhibition.
g. To determine the effects of seedling-microbe-soil systems on the available concentrations of phenolic acids in soil solutions.
h. To compare the effects of phenolic acids on seedlings in nutrient and soil culture, and
i. To determine the effects of phenolic acids at various life stages.

2.4 Results and Discussion

2.4.1 Effects and Duration of Effects of Phenolic Acids on Seedlings in Nutrient Culture

The primary effect of phenolic acids on cucumber seedlings, our model bioassay species, appears to be root cell membrane perturbation (Glass 1973, 1974; Glass and Dunlop 1974). This primary effect ultimately leads to a reduction in hydraulic conductivity (i.e., reduced water uptake, decreased water potential and turgor pressure; Einhellig et al. 1985; Einhellig 1986; Booker et al. 1992), a reduction in net nutrient uptake of ammonium, nitrate, potassium, and phosphorous, among others (Balke 1985; Blum and Dalton 1985; Blum et al. 1985a; Lyu and Blum 1990; Holappa and Blum 1991; Bergmark et al. 1992; Booker et al. 1992; Yu and Matsui 1997; Lehman and Blum 1999a, b; Blum and Gerig 2005), an increased efflux of potassium (Bergmark et al. 1992; Booker et al. 1992; Baziramakenga et al. 1995; Yu and Matsui 1997), changed root cell processes such as down-regulation of cell cycle-related genes, endoreduplication and associated blockage of mitosis (Zhang et al. 2009), increased lignification of cell walls and associated modification of enzyme activity (Shann and Blum 1987b; Politycka 1998; Politycka and Mielcarz 2007; Zanardo et al. 2009), and a host of other secondary, tertiary, etc. effects (frequently just called secondary effects). Increased abscisic acid levels (Holappa and Blum 1991), secondary root initiation, and root/shoot ratios (Blum and Rebbeck 1989), and reduced rates of photosynthesis (Einhellig et al. 1970; Patterson 1981; Blum and Rebbeck 1989), carbon allocation to roots (Blum and Rebbeck 1989), transpiration (Blum et al. 1985a; Barkosky and Einhellig 1993), water utilization (Blum et al. 1985a, b; Lyu and Blum 1990; Holappa and Blum 1991; Blum and Gerig 2005), leaf area expansion (Blum and Dalton 1985; Holappa and Blum 1991; Lehman et al. 1994; Blum and Gerig 2005), and root elongation (Blum and Rebbeck 1989; Klein and Blum 1990a; Zhang et al. 2009) are additional secondary effects which appear to be attributable to reduced water and net nutrient uptake. All these secondary effects in turn lead to slower seedling development and reductions in root and shoot biomass (Rice 1984; Blum and Dalton 1985; Blum and Rebbeck 1989). The intensity and duration of primary and secondary effects are dependent on the sensitivity of the seedling (e.g., health, acclimation, and taxon), types of phenolic acids present, and the continued presence of active phenolic acid concentrations surrounding roots (Einhellig and Kuan 1971; Glass and Dunlop 1974; Blum and Dalton 1985; Blum et al. 1985a, b; Blum and Rebbeck 1989; Lehman and Blum 1999a, b; Blum and Gerig 2005). In general early detectable effects of phenolic acids in nutrient culture ranged from seconds for membrane perturbation to 1 or 2 days for leaf expansion (Glass and Dunlop 1974; Blum and Dalton 1985; Blum and Rebbeck 1989). Detectable recovery of processes such as membrane perturbation and leaf expansion after phenolic acids were depleted or removed from the nutrient solution occurred over a similar time frame as that observed for detectable phenolic acid effects (Glass and Dunlop 1974; Blum and Dalton 1985).

It is also possible that internal cell membranes and organelles such as chloroplasts and mitochondria (Moreland and Novitzky 1987) may be impacted directly by phenolic acids since phenolic acids are taken up by roots and translocated throughout cucumber seedlings (Shann and Blum 1987a, b; Lyu et al. 1990; Lehman and Blum 1999b; Blum and Gerig 2005). However, once they contact and/or are taken up by roots the action of phenolic acids can be readily neutralized by sorption to cell walls, incorporation into lignin, glucosylation, metabolism, and/or sequestration in vacuoles (Harborne 1982; Shann and Blum 1987b; Fry 1988; Politycka et al. 2004). Thus concentrations of phenolic acids reaching aboveground tissues of cucumber seedlings tend to be small compared to concentrations observed in root tissues (Shann and Blum 1987b). For example, 8 day old cucumber seedlings growing in nutrient solution and treated with 0.5 mM ferulic acid, pH 5.5, retained 74% of the ferulic acid in root tissues 5 h after treatment. For 18-day old seedlings this value increased to 91%. In fact, the primary effects of phenolic acids tend to be localized to the initial site of root contact since inhibition of cucumber seedling processes was directly related to the percent of their root systems in contact with phenolic acids (Fig. 2.5; Klein and Blum 1990a; Lyu and Blum 1990; Lehman et al. 1994; Lehman and Blum 1999b) and was poorly related to phenolic acid taken up by roots (Lehman and Blum 1999b). There is, therefore, little data to suggest that the effects of phenolic acids are systemic in nature.

Cinnamic acid derivatives as a rule tend to be more inhibitory than their corresponding benzoic acid derivatives (caffeic, ferulic, *p*-coumaric, sinapic acids vs. protochatechuic, vanillic, *p*-hydroxybenzoic, syringic, respectively). The relative potencies of ferulic, *p*-coumaric, vanillic, sinapic, syringic, caffeic, *p*-hydroxybenzoic and protochatechuic acids based on regression slopes of leaf area for cucumber seedlings growing in a nutrient culture with an initial pH of 5.8 and a range of concentrations from 0 to 1 mM were 1.00, 0.86, 0.74, 0.68, 0.66, 0.65, 0.50,

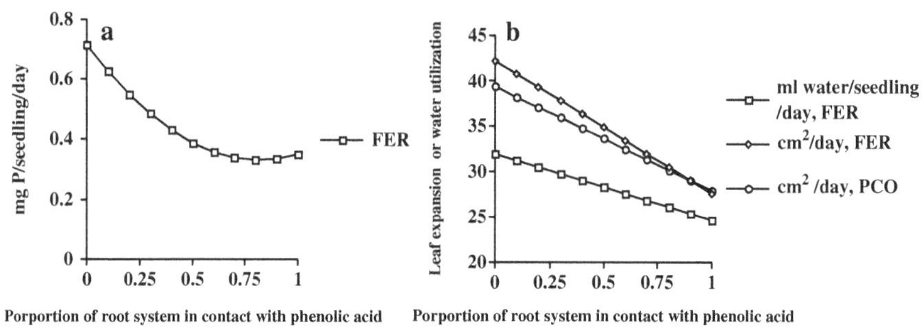

Fig. 2.5 Changes in net phosphorous uptake (**a**; $r^2 = 0.52$), net water uptake (**b**; $r^2 = 0.19$), and absolute growth rates of leaf expansion (**b**; r^2 for FER = 0.76 and PCO = 0.58) of 13–15 day-old cucumber seedlings as the proportion of the root systems in contact with a phenolic acid was increased in nutrient culture, where FER equals 0.5 mM ferulic acid and PCO equals 0.5 mM *p*-coumaric acid. Figures based on regressions from Lyu and Blum (1990) (**a, b**) and Lehman et al. (1994) (**b**). Plenum Publishing Corporation, regressions used with permission of Springer Science and Business Media

and 0.35, respectively (Blum et al. 1985a). The magnitude of observed effects for any given phenolic acid, however, was determined by the concentration of a given phenolic acid and the pH of the nutrient solutions in which cucumber seedlings were growing (Fig. 2.6; Blum et al. 1985b; Lehman and Blum 1999b). Effects of phenolic acids occur primarily under acidic conditions since the pK_a of simple phenolic acids is approximately 4.5 (Fig. 2.7; Blum et al. 1985b, 1999b). Under neutral or basic conditions simple phenolic acids such as ferulic acid and *p*-coumaric acid have a negative charge (are ionized) and are thus not thought to interact or to be taken up by roots which also tend to have negatively charged surfaces (Harper and Balke 1981; Blum et al. 1985b; Shann and Blum 1987a; Lehman and Blum 1999b). According to this hypothesis the contact and/or uptake of phenolic acids by roots requires a protonated state, which predominately occurs under acidic conditions. Recently, Ohno et al. (2002) proposed an additional mechanism. They propose that decreasing pH protonates the functional –OH groups on cell walls to form positively charged $-OH_2^+$ groups which then bind the negatively charged phenolic acids. The increasing electrostatic binding of phenolic acids to cell walls with decreasing pH then directly or indirectly leads to an increase in phytotoxicity by modifying the composition of the solution reaching the cell membranes and/or by modifying cell wall chemistry. The mechanism by which these electrostatically bound phenolic acids may influence the phytotoxicity of phenolic acids is presently not known (Ohno 2009, personal communication). However, no matter the mechanism the effects of phenolic acids on seedling processes can range from stimulation to inhibition depending on the

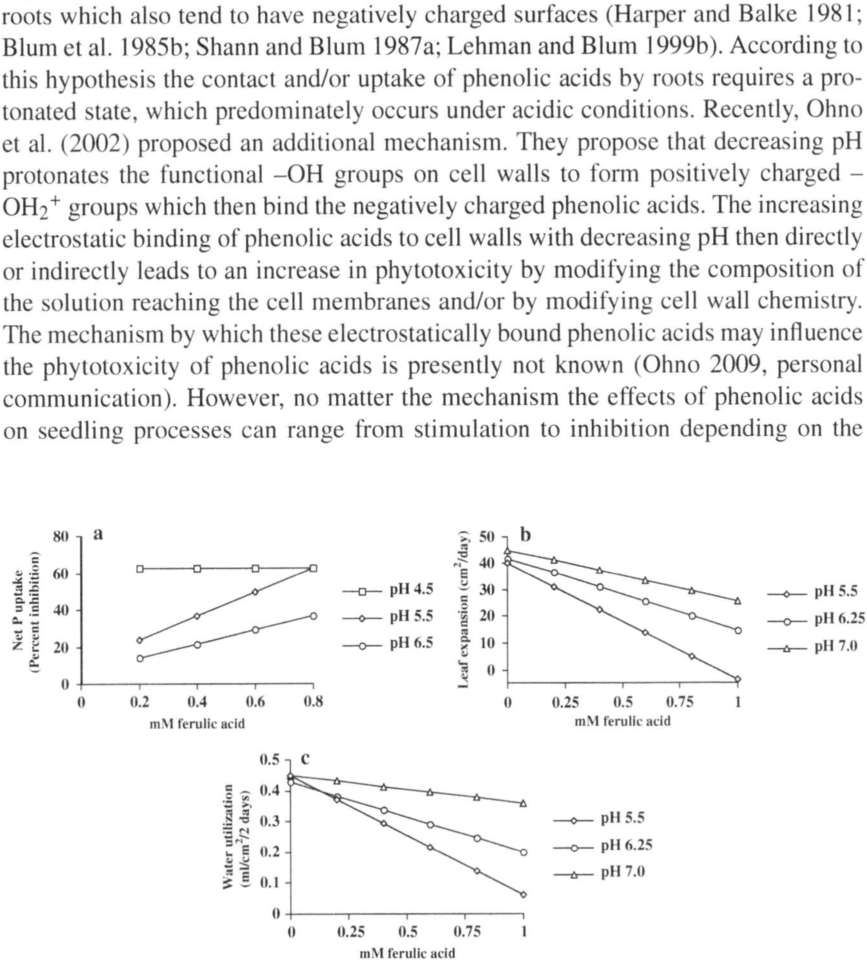

Fig. 2.6 Effects of ferulic acid and initial nutrient solution pH on net phosphorous uptake (**a**; 22 day old; r^2 for pH 5.5 = 0.71, and pH 6.5 = 0.45), absolute growth rates of leaf expansion (**b**; 16–18 day old; r^2 for pH 5.5 = 0.90, pH 6.25 = 0.69, and pH 7.0 = 0.72), and net water utilization (**c**; 16–18 day old; r^2 for pH 5.5 = 0.95, for pH 6.25 = 0.88, and for pH 7.0 = 0.69) of cucumber seedlings. Figures based on regressions and data from Lehman and Blum (1999b) (**a**) and regressions from Blum et al. (1985b) (**b**, **c**). Plenum Publishing Corporation, regressions and data used with permission of Springer Science and Business Media

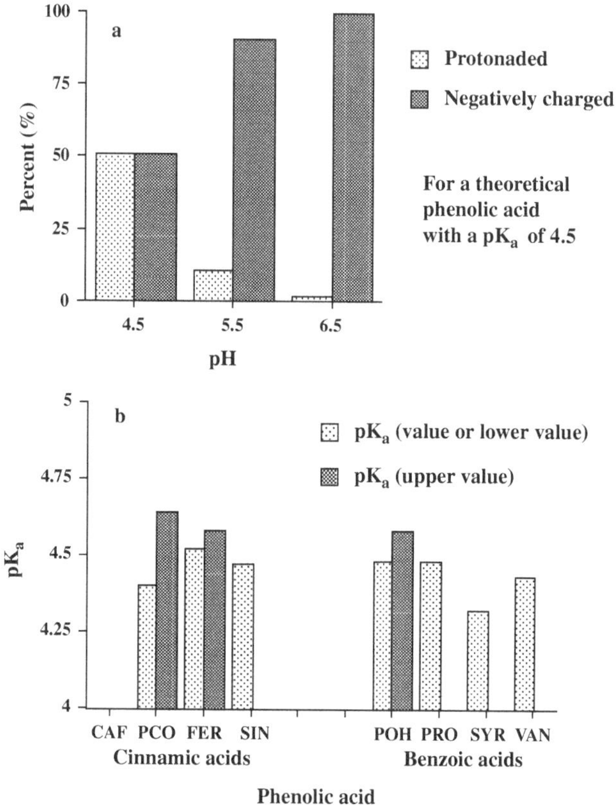

Fig. 2.7 The effects of pH on the ionic state of a theoretical phenolic acid with a pK_a value of 4.5 (**a**) and estimated pK_a values for cinnamic and benzoic acids (**b**). Where CAF equals caffeic acid, PCO equals *p*-coumaric acid, FER equals ferulic acid, SIN equals sinapic acid, POH equals *p*-hydroxybenzoic acid, SYR equals syringic acid, and VAN equals vanillic acid. A pK_a value for caffeic acid was not available. Figure (**b**) based on data from Blum et al. (1999b). CRC Press LLC, data used with permission of Taylor & Francis Ltd, http://www.tandf.co.uk/journals. Original sources of data: AJ Leo, personal communication, Leo et al. (1971), Nordstrom and Lindberg (1965), Kenttamaa et al. (1970), Connors and Lipari (1976); Glass (1975)

concentration and physicochemical state of the phenolic acids (Rice 1986; Pandey 1994; Duke et al. 2006; Belz et al. 2007; Belz 2008).

In addition to type of phenolic acid, phenolic acid concentration, and solution pH there are other factors that influence cucumber seedling responses to phenolic acids. For example:

a. As the number of phenolic acids was increased in a mixture, the concentrations of the individual phenolic acids required to bring about a given inhibition declined (Blum et al. 1985a; Lyu et al. 1990; Lehman et al. 1994; Blum 1996).

b. Effects of individual phenolic acids in a mixture were found to be either additive (lower concentrations) or partially antagonistic (higher concentrations) (Blum et al. 1985a; Lyu et al. 1990; Lehman et al. 1994; Blum 1996; Einhellig 1996; Inderjit et al. 2002).
c. The magnitude of the effects of phenolic acids or phenolic acid mixtures was directly proportional to the root system in contact with phenolic acids (Fig. 2.5; Klein and Blum 1990a; Lyu and Blum 1990; Lehman et al. 1994; Lehman and Blum 1999b).
d. Subjecting cucumber seedlings to stress (e.g., inhibitory ferulic acid concentrations, drought, or declining nutrient supply) prior to ferulic acid treatment reduced, and in some instances eliminated, the inhibition of subsequent ferulic acid treatments (Lehman and Blum 1999a). However, high and low temperatures and drought stress during phenolic acid exposure either enhanced or reduced allelopathic effects (Glass 1976; Duke et al. 1983; Einhellig and Eckrich 1984; Einhellig 1987, 1996, 1999), and
e. The ability of cucumber seedlings to modify available/active phenolic acid concentrations surrounding their roots (e.g., sorption to cell walls, uptake and neutralization, and rhizosphere pH changes) indicated that cucumber seedlings themselves directly influenced the magnitude of primary and secondary effects of phenolic acids through feedback regulation (Blum and Gerig 2005).

There were also factors that have limited influence or no influence on the effects of phenolic acids. For example:

a. Inhibition of radicle growth by phenolic acids over the first 48 h of germination did not influence the subsequent growth of seedlings. Thus the effects of phenolic acids were simply to delay the germination process (Blum et al. 1984; Blum and Dalton 1985), and
b. Pretreatment of cucumber seedlings with non-inhibitory concentrations of phenolic acids did not modify the subsequent effects of inhibitory concentrations (Blum and Dalton 1985).

Finally, once phenolic acids were removed from the root environment, either by seedling uptake, microbial utilization, and/or physical removal (e.g., replacing nutrient solutions), seedling processes (e.g., leaf area expansion, transpiration, root growth, nutrient uptake, and water uptake) recovered (Blum and Dalton 1985; Blum et al. 1985a; Blum and Rebbeck 1989; Bergmark et al. 1992; Blum and Gerig 2005). Recovery was faster for seedlings grown in an adequate nutrient environment than for seedlings grown in a limited nutrient environment (Blum and Dalton 1985). Anatomical and morphological characteristics that were fixed, however, did not change. For example, once a leaf is fully expanded, no further change in leaf area will occur unless the leaf wilts. However, growth of new and expanding leaves will recover as phenolic acid concentrations decline although the final leaf area may still be reduced (Blum and Gerig 2005). Absolute and relative rates of cucumber

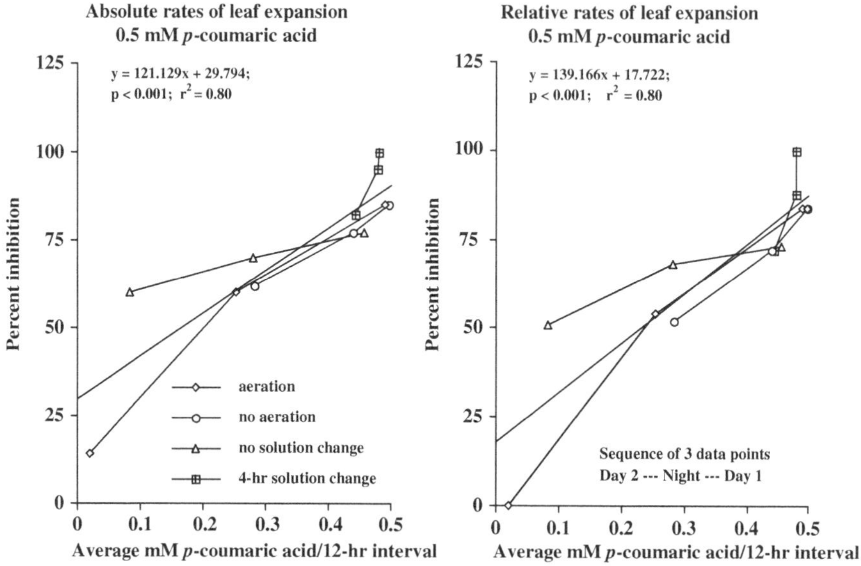

Fig. 2.8 Change in absolute and relative rates of leaf expansion of 12 day-old cucumber seedlings as *p*-coumaric acid declines due to root uptake and microbial utilization in nutrient culture in the presence and absence of aeration, and when solutions were not changed or changed every 4 h. Figures reproduced from Blum and Gerig (2005). Figures used with permission of Springer Science and Business Media

seedling leaf expansion, for example, increased 12 and 14%, respectively, for every 0.1 mM decline in *p*-coumaric acid (Fig. 2.8; Blum and Gerig 2005).

In summary, primary effects (i.e., membrane perturbations) of phenolic acids reduce water and net nutrient uptake and these reductions lead to a host of secondary effects (e.g., inhibition of photosynthesis, carbon allocation, and growth). There is little data to suggest that effects of phenolic acids are systemic; the primary effects appear to be localized to root contact and the majority of phenolic acids taken up by roots are retained in the roots. Seedling roots can moderate the effects of phenolic acids through regulating available/active phenolic acid concentrations surrounding roots (i.e., feedback regulation). In addition, inhibitory effects of phenolic acids vary with species sensitivity, type of phenolic acid (e.g., cinnamic acids are more potent than corresponding benzoic acids), concentration, pH, composition of phenolic acid mixtures (e.g., required concentrations for individual phenolic acids decline as the number of phenolic acids in a mixture are increased), pre-treatment stress (acclimation), environment during treatment, proportion of roots in contact with phenolic acids (e.g., as proportion increases so does inhibition), and duration of contact. Once phenolic acids are neutralized by microbial utilization, uptake by roots, etc., primary and secondary processes can recover, the better the growth environment the faster the recovery. These observations indicate that the relationships between phenolic acid concentrations and observed seedling effects are extremely complex, essentially a "moving target", depending on the nature, number and concentrations of the phenolic acids present, the physicochemical environment, and status of sensitive seedling (e.g., health, acclimation, etc.).

2.4.2 *Effects of Seedlings, Mixtures of Phenolic Acids, and Microbes on Phenolic Acid Concentrations in Nutrient Culture*

Cucumber seeds and seedlings have associated with them substantial microbial populations that are difficult to eliminate because microbes are not only found on and in the cutinized surface of the seed coat but can also be found internally within the seed (Leben 1961; Mundt and Hinkle 1976). Depletion of phenolic acids from nutrient solutions thus represent uptake by roots and microbial utilization. By replacing the nutrient solution (control) and nutrient-phenolic acid solutions (treatments) every other day, microbial populations were kept in check and phenolic acid concentrations were brought back to the original treatment concentrations. However, since phenolic acid treatments changed microbial populations on the rhizoplane (Fig. 2.9)

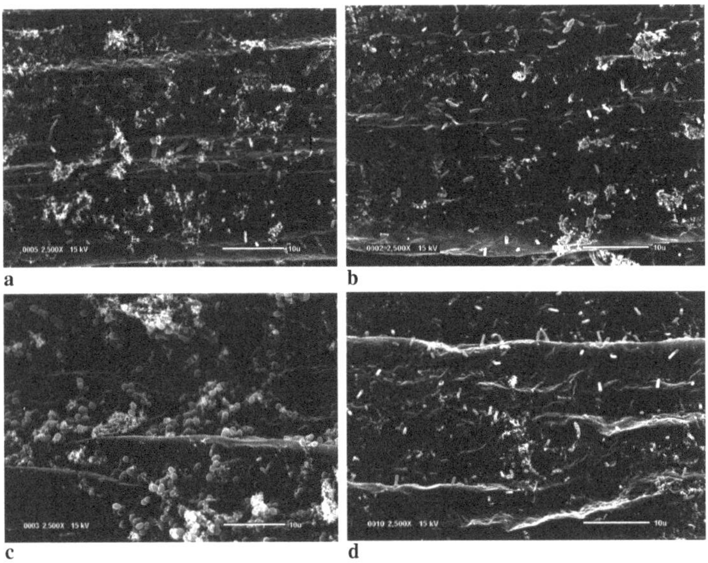

Fig. 2.9 Electron micrographs (2500× 17 kv) of root surfaces of 13 day-old cucumber seedlings grown in nutrient culture not treated (controls; **a**, **b**) or treated 4 times (starting with day 6) every other day with 0.5 mM *p*-coumaric acid (**c**, **d**). Nutrient solutions (pH 5.0) with or without *p*-coumaric acid were changed every other day. Fine matrix material in micrographs is very likely mucigel generated by root and associated microbes. Micrographs chosen represent the maximum (**a**, **c**) and minimum (**b**, **d**) differences observed for 8 micrographs taken along the first 10 mm (tip) of the control and *p*-coumaric acid treated roots. Finally, microbes observed in these micrographs represent all types of microbes, not just microbes that can utilize phenolic acids as a sole carbon source, since phenolic acid utilizers cannot be distinguished by morphology from other carbon utilizers

and in the nutrient solution (solutions became cloudy over time), the best quantitative estimates of root uptake based on solution depletion were thus just after phenolic acid solutions were changed. Initial phenolic acid microbial breakdown products were detectable within 8–10 h after solution change (Blum et al. 1985a; Shann and Blum 1987a; Blum and Gerig 2005).

Depletion of ferulic acid/g root at the end of a 5 h treatment period increased in a curvilinear manner between 0.1 and 0.7 mM, for 12-day old cucumber seedlings (Fig. 2.10; Lehman and Blum 1999b). Depletion of ferulic acid by 14- to 18-day old cucumber seedlings at the end of a 5 h treatment period was also curvilinear and decreased with increasing solution pH (Fig. 2.10), but depending on treatment concentration, ranged from linear to curvilinear during the 5 h treatment period (Shann and Blum 1987a). A 25% depletion of *p*-coumaric acid from nutrient culture (initial solution pH 5.0 and solution volume of 110 ml) by 13-day old cucumber seedlings was estimated to take approximately 8 h for 0.125 mM, 11–12 h for 0.25 mM, 18–19 h for 0.5 mM, 23–25 h for 0.75 mM and 28–29 h for 1 mM treatments (Blum and Gerig 2005).

The type of relationship observed (i.e., significant or not significant) between depletion rates and *p*-coumaric acid concentrations in the nutrient solution, and the overall depletion rates, however, varied with time of day and time of day when seedlings were treated with *p*-coumaric acid (Blum and Gerig 2005). When significantly related, depletion of *p*-coumaric acid over the concentration range used increased linearly with increasing concentrations of *p*-coumaric acid. Significant relationships were observed after each *p*-coumaric acid treatment, but also at one other time. Depletion of *p*-coumaric acid occurred at all times; only the rates varied during each period monitored. The observed patterns suggested that periods with significant positive relationships between *p*-coumaric acid depletion and *p*-coumaric

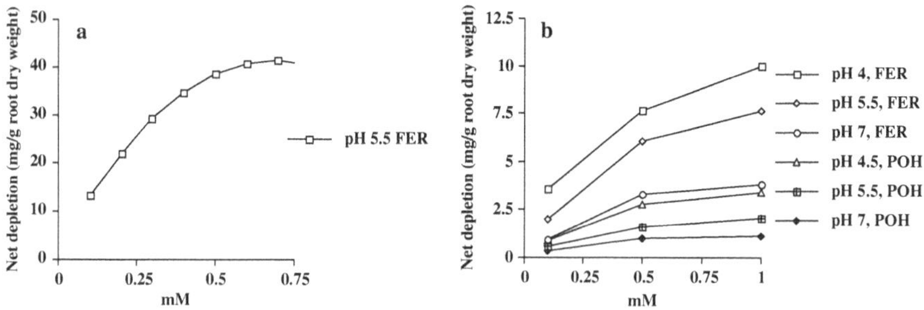

Fig. 2.10 Net depletion of phenolic acid by 12 day-old cucumber seedlings grown in a growth chamber (**a**; $r^2 = 0.78$) and by 14–18 day-old cucumber seedlings grown in a light bank (**b**; $r^2 \geq 0.79$), where FER equals ferulic acid and POH equals *p*-hydroxybenzoic acid. Nutrient solutions were aerated. Initial pH values for nutrient solutions of (**a**) were 5.5. Initial pH values for (**b**) varied as indicated. All phenolic acid values were determined after 5 h. **a** based on regression from Lehman and Blum (1999b) (Plenum Publishing Corporation, regression used with permission of Springer Science and Business Media) and **b** based on data points of two figures from Shann and Blum (1987a)

acid concentrations represented periods dominated by apoplastic movement of *p*-coumaric acid into the intercellular spaces of roots and that periods lacking significant relationships represent periods dominated by symplastic uptake and movement. Average 24-h depletion rates for 13- to 15-day old cucumber seedlings ranged from about 0.02–0.04 mM/g root dry weight/h for *p*-coumaric acid.

In addition, depletion rates also varied with the type of phenolic acid present, the number of phenolic acids treatments received by seedlings, the number and concentrations of phenolic acids present in a solution, and the level of aeration of the nutrient solution (Blum and Dalton 1985; Blum et al. 1985a; Shann and Blum 1987a; Lyu et al. 1990; Lyu and Blum 1990; Lehman and Blum 1999b; Blum and Gerig 2005). For example:

a. Shann and Blum (1987a) observed that depletion of 0.1–1 mM ferulic acid by 14- to 18-day old cucumber seedlings was 50% to 75% (1.5–1.75 times) higher than the depletion of 0.1–1 mM *p*-hydroxybenzoic acid. Lyu et al. (1990) observed that depletion of ferulic acids (0.125–0.5 mM) for 15-day old cucumber seedlings was 1.5–2 times greater than vanillic acid and 1.3–2 times greater than *p*-coumaric acid.
b. Uptake of ^{14}C-labeled ferulic acid by cucumber seedling roots treated once to 0.5 M ferulic acid was 50% (1.5 times) higher than seedlings treated twice to ferulic acid, 48 h apart (Shann and Blum 1987a). Depletion of ferulic acid and ^{14}C uptake from nutrient solution were consistently related, but depletion rates were 2.5 times greater than ^{14}C uptake.
c. The depletion of a range of concentrations (0.25–1 mM) of ferulic acid by cucumber seedlings was not modified by the presence of 0.25 or 0.5 mM *p*-hydroxybenzoic acid, but the depletion of *p*-hydroxybenzoic acid was reduced by 30% in the presence of ferulic acid (Shann and Blum 1987a). Lyu et al. (1990) found similar behavior for two-way equal-molar combinations of 0.125–0.5 mM ferulic acid, vanillic acid and *p*-coumaric acid. However, in their study ferulic acid depletion by cucumber seedlings was also depressed. The magnitude of depression for each of the phenolic acids in the presence of another phenolic acid varied with concentration ranging from 22 to 82% (Fig. 2.11), and
d. Depletion of phenolic acids was much greater with aeration than without aeration (Blum and Gerig 2005). A 25% depletion of 0.5 mM *p*-coumaric acid took 14.5 h for the aerated nutrient solution and 24 h for the non-aerated nutrient solution (Fig. 2.12). In addition, there was a greater accumulation of initial phenolic acid breakdown products without aeration than with aeration. Breakdown products of *p*-coumaric acid in *p*-coumaric equivalence increased over time in the absence of aeration and increased ever so slightly before decreasing in the presence of aeration (Fig. 2.12). An initial microbial breakdown product of *p*-coumaric acid is *p*-hydroxybenzoic acid (Blum 1998). In fact, initial microbial breakdown products of simple phenolic acids are frequently other phenolic acids, for example, ferulic acid is converted by fungi to caffeic acid or vanillic acid, and these in turn are converted to protocatechuic acid (Evans 1963; Dagley 1971; Martin and Haider 1971, 1976).

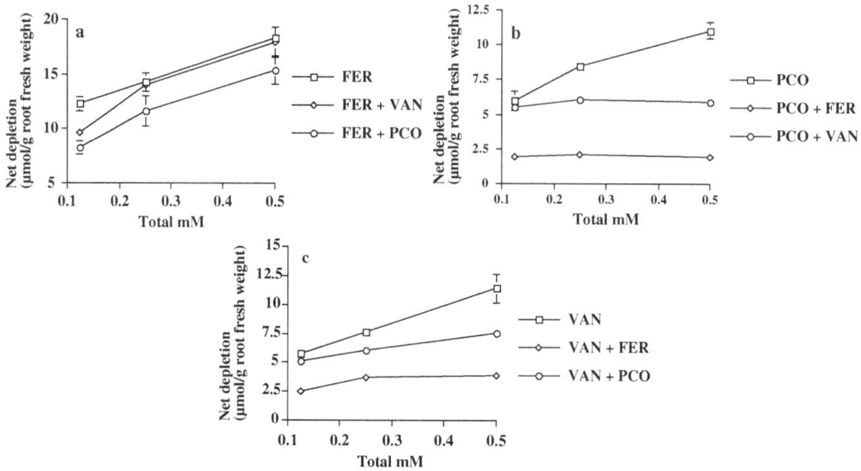

Fig. 2.11 The net depletion of phenolic acids in the absence or presence of a second phenolic acid at equal-molar concentrations from nutrient solution by 15-day old cucumber seedlings growing in a light bank. Where FER equals ferulic acid, PCO equals *p*-coumaric acid, and VAN equals vanillic acid and data in (**a**) are depletion of ferulic acid, **b** depletion for *p*-coumaric acid, and **c** depletion for vanillic acid. Nutrient solutions were not aerated and had an initial pH of 5.5. The absence of standard error bars indicates that the error bars are smaller than the symbols representing the mean. Figures based on data from Lyu et al. (1990). Plenum Publishing Corporation, data used with permission of Springer Science and Business Media

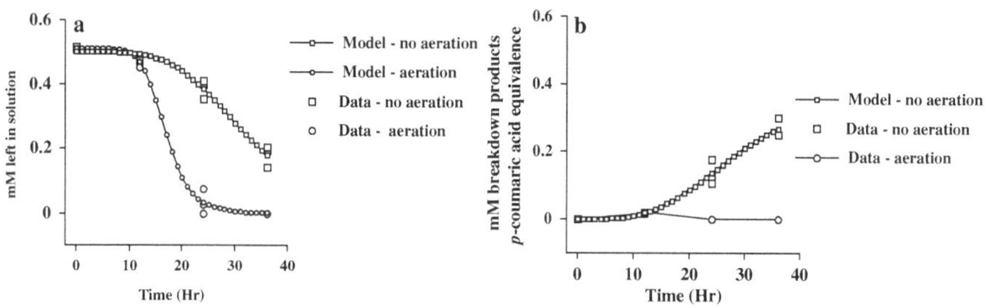

Fig. 2.12 The decline of 0.5 mM *p*-coumaric acid (**a**) and the accumulation and decline of initial phenolic acid breakdown products (**b**) in nutrient solutions (pH 5.0) surrounding roots of 12 day-old cucumber seedlings. Breakdown products are in *p*-coumaric acid equivalence. Nutrient solutions were aerated or not aerated. Figures reproduced from Blum and Gerig (2005). Figures used with permission of Springer Science and Business Media

Since depletion is closely related to root uptake in these nutrient culture studies (Shann and Blum 1987a), some comments on the distribution of phenolic acids in seedlings are warranted. Once phenolic acids contact and/or are taken up by roots their action can readily be neutralized by sorption to cell walls, incorporated into lignin, glucosylation, metabolism, and/or sequestration in vacuoles (Shann and Blum 1987b; Fry 1988; Harborne 1982). Thus concentrations of phenolic acids reaching aboveground tissues of cucumber seedlings tend to be small compared to concentrations observed in root tissues (Shann and Blum 1987b). For example, 8 day old cucumber seedlings growing in nutrient solution and treated with 0.5 mM ferulic acids, pH 5.5, retained 74% of the ferulic acid in root tissues 5 h after treatment. For 18-day old seedlings this value increased to 91%.

In summary the rate of depletion (i.e., root uptake and microbial utilization) varies with type phenolic acid present, concentration, pH, time of day, time of day of treatment, number of treatments, composition of phenolic acid mixtures, whether uptake is apoplastic or symplastic, phenolic acid-utilizing microbial populations present on roots and in the nutrient solution, and aeration. Phenolic acid treatments of seedlings in nutrient culture modify microbial populations on root surfaces (rhizoplane) and in the nutrient solutions. Once taken up by roots, phenolic acids were distributed throughout seedlings. Highest concentrations, however, were retained in the roots.

2.4.3 *Interactions of Phenolic Acids with Sterile and Non-sterile Soils*

When simple phenolic acids (e.g., cinnamic or benzoic acids) enter the soil environment a number of processes occur to modify the available/active fractions of phenolic acids in the soil. The available fraction is composed of "free" phenolic acids in the soil solution and reversibly sorbed phenolic acids on soil particles. The active fraction constitutes that portion of available phenolic acids in the soil which actually interacts with roots and microorganisms located in the rhizosphere and on the rhizoplane. Processes influencing the available/active fractions include irreversible sorption to soil particles (e.g., bound into recalcitrant organic matter or onto clay minerals), polymerization, oxidation-reduction, leaching, microbial utilization, root uptake, and soil pH (Shindo and Kuwatsuka 1976; Huang et al. 1977, 1999; Dao 1987; Lehmann et al. 1987; Pue et al. 1995; Makino et al. 1996; Dalton 1999; Schmidt and Ley 1999; Blum et al. 1999b; Ohno 2001; Blum 2004, 2006; Inderjit and Bhowmik 2004; Inderjit 2005; Tharayil et al. 2006). Since the magnitude of plant–plant allelopathic interactions is in part determined by the concentration of the active fraction of phenolic acids in the soil (other factors include seedling sensitivity, state of acclimation, environmental stresses, etc.), obtaining reliable quantitative estimates of the active fraction of phenolic acids is essential if we are to identify and characterize the role of phenolic acids in plant–plant allelopathic interactions. However, because of the difficulty of actually determining the active fraction, such data are not presently available. The best we can do is to estimate the available fraction. The available fraction (phenolic acids in soil solutions and reversibly sorbed to soil particles) includes the active fraction, but these two fractions may or may not be consistently and/or directly related to each other.

2.4.3.1 Extraction Procedures

Various soil extraction procedures utilizing a range of inorganic and organic solvents, extraction times, and soil sterilization procedures have been used to extract phenolic acids from soils (Guenzi and McCalla 1966; Turner and Rice 1975; Chou and Patrick 1976; Kaminsky and Muller 1977, 1978; Lindsay and Norwell 1978; Caballeira and Cuervo 1980; Katase 1981b; Whitehead et al. 1982; Cheng et al. 1983; Dalton et al. 1983, 1987, 1989a, b; Dalton 1989, 1999; Blum et al. 1991, 1994, 1999b; Blum 1997, 1998). As might be expected, the recovery of phenolic acids from soil varied with the type of phenolic acid, soil type, soil horizon, extraction procedure, nutrient level, size of phenolic acid-utilizing microbial populations, and time. In general, water extractions recovered the least and NaOH recovered the most (Fig. 2.13 and Table 2.3; also see Whitehead et al. 1983; Dalton 1999). What was also clearly evident from the data was that a number of these extraction procedures (e.g., various NaOH extractions) recovered fractions of phenolic acids from soil organic matter or clay particles that would not be involved in plant–plant allelopathic interactions (Kaminsky and Muller 1977, 1978; Dalton et al. 1987; Dalton 1999; Blum et al. 1994). Thus the critical question for researchers studying plant–plant allelopathic interactions: Which of these extraction procedures actually recover available (i.e., biologically important) phenolic acids from soils, i.e., phenolic acids in soil solution and reversibly sorbed to soil particles that can potentially interact with roots and microbes. The literature suggested that water extractions recover the "free" phenolic acids in soil solution (Shindo and Kuwatsuka 1976; Whitehead et al. 1983) and extractions utilizing mild extractants (e.g., chelates such as neutral EDTA or DTPA, Na acetate, citrate, and Mehlich III) recovery available phenolic acids in soil (Martin and Haider 1976; Kaminsky and Muller 1977, 1978; Dalton et al. 1983, 1989; Dalton 1989a, b; Blum et al. 1994). We settled on water and neutral disodium ethylenediaminetetraacetic acid (EDTA, pH 7) to estimate available phenolic acids in our soils (Blum et al. 1994). We found that 0.25–0.5 M EDTA

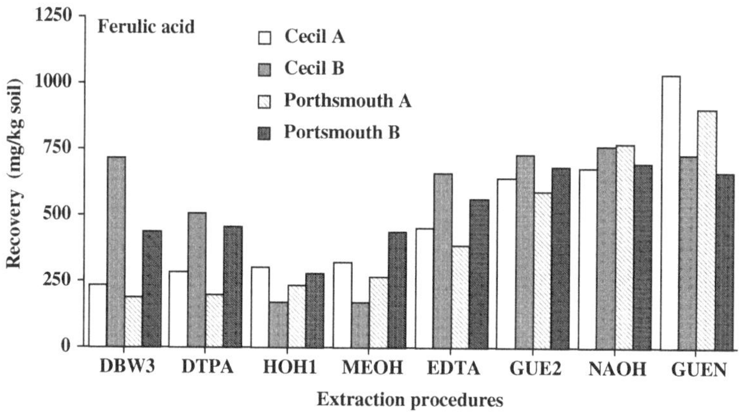

Fig. 2.13 Recovery of ferulic acid by various extraction procedures from sterile soils 90 days after ferulic acid solutions (1,000 mg/kg soil, pH 6.0) were added to soils. Soil-ferulic acid mixtures were stored in the dark at room temperature. $LSD_{0.05}$ for Cecil A and B and Portsmouth A and B soils were 28.70, 44.15, 40.69, and 28.66, respectively. Meaning of the abbreviations and details for extraction procedures are provided in Table 2.3. Figure based on data from Dalton et al. (1987). Data used with permission of Soil Science Society of America

Table 2.3 Details for extraction procedures for Fig. 2.13[a]

Abbreviations	Extractant	Soil/extractant ratio	Concentration	pH	Extraction time	Temperature (°C)
DBW3	Na_2EDTA[b,c]	1:100	0.03 M	7.5	30 min	Room temp.
DTPA	DTPA[b,c]	1:2	0.005 M	7.3	2 h	25
HOH1	H_2O	1:43	100%	6–7	Immediate	22–23
MEOH	MeoH[c]	1:43	100%	–	48 h	10
EDTA	EDTA	1:4	0.5 M	7.5	5 h	Room temp.
GUE2	NaOH	1:1	2 M	14	4 h	Room temp.
NAOH	NaOH	1:25	2 M	14.3–14.7	24 h	Room temp.
GUEN	NaOH	1:43	2 M	14	45 min	121[d]

[a]Source of information: Dalton et al. (1987), information used with permission of Soil Science Society of America. Reference sources for these extraction procedures can be found at Dalton et al. (1987).
[b]EDTA = etheylenediaminetetratacetic acid, DTPA = diethylenetriaminepentaacetic acid.
[c]Constant mixing during extraction.
[d]Autoclaved.

(pH 7) with an extraction time of 2–5 h was the most satisfactory for our soils. It was, however, essential that the pH of the EDTA-soil slurry was maintained at pH 7 over the entire extraction period.

Justifications for using water and neutral EDTA were as follows: To obtain an estimate of "free" phenolic acids in soil we concluded, based on the extraction procedures utilized, the literature, and the fact that water is an essential component/solvent of soil solutions that water extractions would provided the most reasonable estimate of "free" phenolic acids in soils (Whitehead et al. 1983; Dalton et al. 1987; Dalton 1999; Blum et al. 1994; Strobel 2001). The use of neutral EDTA as the primary extractant to estimate available phenolic acids in our soils was based on the following observations. In the presence of phenolic acid-utilizing microbes neutral EDTA extractions of soils eventually cease to recover phenolic acids because of the depletion of soil solution and reversibly sorbed phenolic acids by sorption and microbial utilization and the inability of EDTA to extract more tightly sorbed phenolic acids (Blum et al. 1994). In addition EDTA extractions are ineffective in extracting tightly sorbed phenolic acids from aged plant residues, a potential future source of phenolic acids released by microbial decomposition and leaching of plant residues (Blum et al. 1992).

The actual effectiveness of neutral EDTA as an extractant of available phenolic acids was demonstrated as follows (Blum et al. 1994):

a. Sterilized Cecil A and B soils were amended with 2.5 μmol/g soil of sterile phenolic acid solutions, or sterile water. The soil samples were allowed to equilibrate for 50 days before being inoculated with microbes that could utilize phenolic acids as a carbon source. EDTA extractions of soil samples were made at various time intervals until the 0.25 M EDTA (pH 7, extraction time 5 h) extractions recovered less than 0.01 μmol/g soil phenolic acid. For ferulic acid, *p*-hydroxybenzoic acid, *p*-coumaric acid, and vanillic acid that amounted to < 2 μg/g soil, and
b. Soils were then extracted with 1 M NaOH (pH 12.5, extraction time 5 h). Since 1 M NaOH extractions can recover some irreversibly sorbed phenolic acids the amounts recovered that were > 0.01 μmol/g soil were therefore irreversibly sorbed phenolic acids not available to microbes and presumably roots.

Recovery of phenolic acids by NaOH from amended soil samples was equal to or greater than those recovered from non-amended soil samples. The difference (minus approximately 0.01 μmol/g soil) in recovery between amended and non-amended phenolic acid soils represented a portion of the amended phenolic acids (i.e., 1 M NaOH extractable) that had been irreversibly sorbed during the equilibration and/or incubation periods. Values for 1 M NaOH extractable phenolic acids from non-amended soils were < 0.0017 μmol/g soil for Cecil B (0.2% organic matter) soil samples and ranged from 0.013 μmol/g soil for ferulic acid to 0.073 μmol/g soil for *p*-coumaric acid in the Cecil A (3.7% organic matter) soil samples. Differences between amended and non-amended soils ranged from 0 μmol/g soil for *p*-coumaric acid to 0.024 μmol/g soil for ferulic acid in Cecil A soil samples and 0.013 μmol/g soil for *p*-coumaric acid to 0.024 µmol/g soil for ferulic acid in Cecil B soil samples. Based on these observations it was concluded that neutral EDTA was an excellent extractant to estimate available (i.e., soil solution ("free") and reversibly sorbed) phenolic acids in our soils (Blum et al. 1994).

Together water and neutral EDTA extractions provide the following (water and EDTA extractions were made on different soil samples):

a. Water extractions provide an estimate of soil solution concentrations (i.e., "free" phenolic acids). Some very weakly sorbed phenolic acids may also be desorbed by water extraction and included in water extracts (Strobel 2001).
b. EDTA (0.25–0.5 M EDTA; pH 7) extractions provide an estimate of both "free" and reversibly sorbed phenolic acids (i.e., available phenolic acids that can interact with roots and associated microbes), and
c. The difference between water and EDTA extractions provides an estimate of reversibly sorbed phenolic acid concentrations.

Since the water content and solutions added to soil samples was determined and known, respectively, "free", reversibly sorbed, and available phenolic acids of a soil sample could be expressed in either µM or µmol/g soil. In addition since under laboratory conditions the concentrations of phenolic acids added to sterile soils were known, the amount (% or µmol/g soil) of irreversibly sorbed phenolic acids (e.g., bound into recalcitrant organic matter or onto clay minerals) could also be estimated.

Cecchi et al. (2004), however, classified EDTA, ethyl acetate, methanol, or Mehlich III solutions as harsh extractants that appear to be helpful in determining irreversible sorption of phenolic acids but not bioavailable phenolic acids. I would strongly disagree particularly in regard to neutral EDTA extractions. Our data clearly indicate that this is not the case (see justification above). As a matter of fact it is known that roots and microbes in soil produce a number of chelating agents that function in a similar manner to EDTA (Nagarajah et al. 1970; Doetsch and Cook 1973; Kaminsky and Muller 1977; Marschner and Römheld 1996; Fisher 2002; Fisher and Bipp 2002).

2.4.3.2 Sorption

The pK_a values for the cinnamic and benzoic acid derivatives ranges from 4.3 to 4.6 (Fig. 2.7; Blum et al. 1999b), thus both anionic and protonated (nonionic) forms are present in soils. The anionic forms for example can bind with positively charged sites (e.g., Al and Fe-oxyhydroxides) on soil surfaces (Watson et al. 1973; Parfitt et al. 1977) and indirectly to negatively charged sites (e.g., clays) by multivalent cation bridges (Greenland 1965, 1971). The protonated nonionic forms can be sorbed by soil organic matter (Chiou 1989; Hasset and Banwart 1989) and/or polymerized into the humic substances in the soil (Martin and Haider 1971, 1976, 1979; Martin et al. 1972; Haider and Martin 1975; Haider et al. 1977; Wang et al. 1978, 1986). The binding strengths of the resulting complexes vary. Some complexes (e.g., those resulting from anion exchange or cation bridging) are easily disrupted and can reenter the soil solution; others (e.g., those resulting from ligand exchange, oxidative reactions with mineral surfaces, and incorporation into humic substances) are sorbed or bound into recalcitrant organic matter and thus not available to reenter the soil solution, at least not in the short term. Thus extractants with different chemical and physical properties will recover different fractions (e.g., "free", reversibly sorbed, and/or irreversibly sorbed) of phenolic acids from soils. For example Dalton (1999) observed that recovery of ferulic acid 90 days after addition to sterile Cecil and Portsmouth soils was generally in the following order for these extractants: water = methanol < sodium acetate < EDTA = DTPA < sodium hydroxide.

However, sorption/desorption is not only limited to inorganic and organic substances in the soil. For example, at typical soil pH values (pH 5–8) both surfaces of soil and surfaces of soil microorganisms are predominately negatively charged and can go through similar sorption/desorption processes (Hartel 1998). This is also true for primary root surfaces since substances in cell walls such as pectin and many xylans and arabinogalactans contain uronic acid residues making cell walls predominately acidic and negatively charged (Fry 1988).

2.4.3.3 Recovery from Model Soil Systems

Recovery of available fractions by neutral EDTA from our sterilized soils (i.e., Cecil, Portsmouth and/or White Store soils), when soils and phenolic acids were roughly at steady state, were directly related to the concentration of phenolic acid added to soil, inversely related to soil organic matter and soil pH, greater for B-horizon soils than A-horizon soils, and greater for benzoic acid derivatives than cinnamic acid derivatives (Figs. 2.14 and 2.15; Dalton et al. 1983, 1987; Blum et al. 1994). Recovery of phenolic acids by EDTA extraction (i.e., available: "free" and reversibly sorbed phenolic acids) was essentially parallel to that of water extractions (i.e., "free" phenolic acids), but higher (Blum et al. 1994). Reversibly sorbed phenolic acid concentrations (difference between water and EDTA extractions) rapidly increased (sorption) or decreased (desorption) with increasing or decreasing concentrations of the phenolic acids in soil solutions, respectively (Figs. 2.16 and 2.17), and increased as multivalent cations (e.g., Ca^{++}, multivalent cation bridging) increased in the soil (Dalton et al. 1987; Blum 1998; Blum et al. 1999b). Such observations support the conclusion that reversibly sorbed phenolic acids in soils were readily available for utilization by soil microbes and thus also readily available to interact with roots (Blum 1998).

Rates of phenolic acid sorption (reversible and irreversibly sorbed or fixed) to sterile soils varied over time. Immediately after addition of phenolic acids to soils (first 1–4 h) there was a very rapid sorption (primarily irreversible sorption) for both the cinnamic acid and benzoic acid derivatives (Figs. 2.15 and 2.18; Blum et al. 1994, 1999b). Thereafter, irreversible or fixed sorption of cinnamic acid derivatives, but not benzoic acid derivatives, continued at a very much reduced rate. Soil moisture at the time phenolic acids were added to soil also influenced the rates of soil sorption (Blum et al. 1999b). Initial reversible sorption (first 4 h) of ferulic acid in

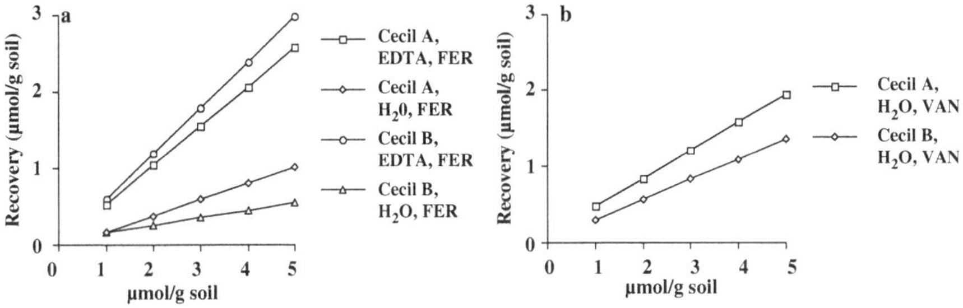

Fig. 2.14 Recovery of ferulic (FER) acid (**a**; $r^2 = 0.99$) and vanillic (VAN) acid (**b**; $r^2 \geq 0.95$) by 0.5 M EDTA (pH 8) or water 42 days after addition of a range of phenolic acid concentrations to sterile Cecil A and B soils. Figures based on regressions from Blum et al. (1994). Plenum Publishing Corporation, regressions used with permission of Springer Science and Business Media

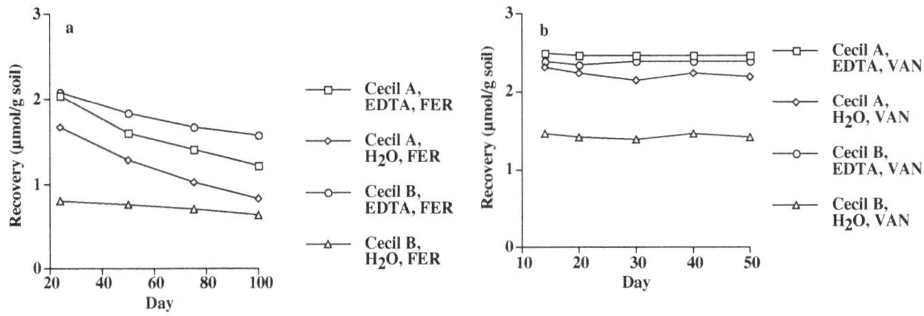

Fig. 2.15 Recovery, over time, of ferulic (FER) acid (**a**; $r^2 \geq 0.89$) and vanillic (VAN) acid (**b**) from sterile Cecil A and B soils by 0.25 M EDTA (pH 7) or water. Phenolic acid added at time zero was 2.5 μmol/g soil. Standard error bars for (**b**) are smaller than the symbol representing the mean. **a** based on regressions and **b** based on data points of two figures from Blum et al. (1994). Plenum Publishing Corporation, regressions and data used with permission of Springer Science and Business Media

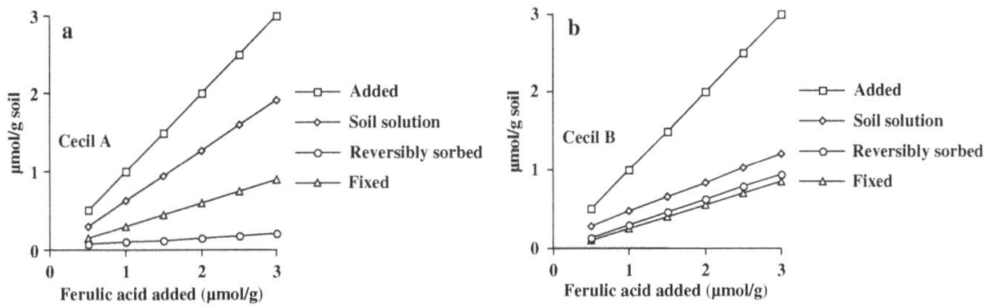

Fig. 2.16 Amounts of ferulic acid in soil solution, reversibly sorbed and fixed (irreversibly sorbed) in sterile Cecil A (**a**) and B (**b**) soils 35 days after addition. Standard error bars for (**a**) and (**b**) are smaller than the symbol representing the mean. Figures reproduced from Blum (1998). Plenum Publishing Corporation, figures used with permission of Springer Science and Business Media

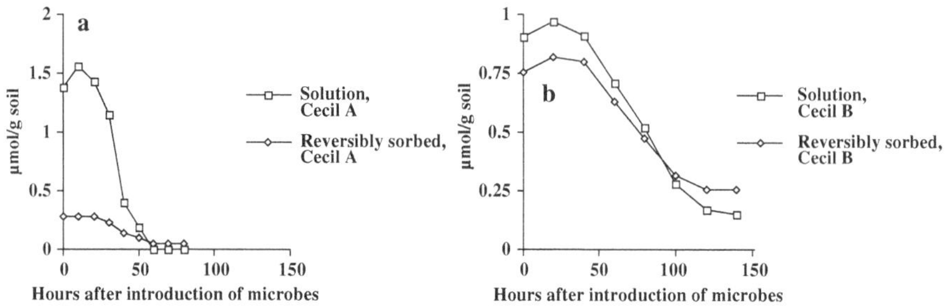

Fig. 2.17 Utilization of ferulic acid in soil solution and reversibly sorbed to Cecil A (**a**) and B (**b**) soils by microbes. Ferulic acid added at time zero was 2 µmol/g soil. Standard error bars for (**a**) and (**b**) are smaller than the symbol representing the mean. Figures reproduced from Blum (1998). Plenum Publishing Corporation, figures used with permission of Springer Science and Business Media

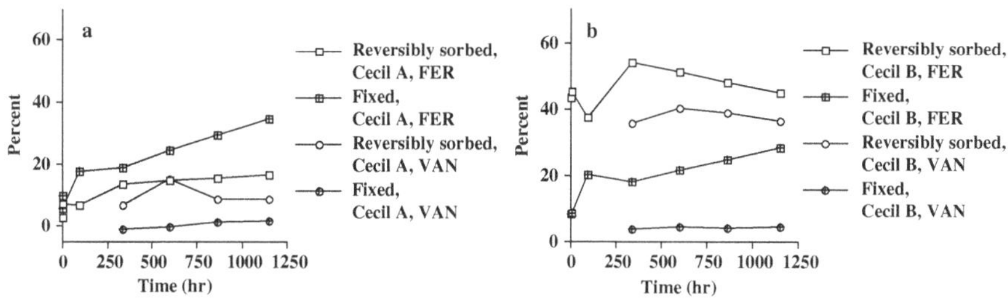

Fig. 2.18 Percent ferulic acid and vanillic acid reversibly sorbed and fixed (irreversibly sorbed) by sterile Cecil A (**a**) and B (**b**) soils over time. Percentages based on 1–3 µmol/g soil added at time zero. Figures based on data from Blum et al. (1999b). CRC Press LLT, data used with permission of Taylor & Francis Ltd, http://www.tandf.co.uk/journals. Original sources of data: Blum (1997, 1998) and Blum et al. (1994)

dry soil was 2.3 and 1.3 times that of wet soils for Cecil A and B, respectively. While irreversible sorption over the same time period was 0.89 and 1.4 times, respectively. Once soils were inoculated with soil microbes available (i.e., "free" and reversibly sorbed) phenolic acids declined rapidly (Fig. 2.17; also see Fig. 2.24), the rate being determined by the size of the phenolic acid-utilizing microbial populations and the soil environment.

Dalton et al. (1989b) utilizing Mehlich III, a mild chelating extractant, also observed that the recovery of available phenolic acids (ferulic acid, vanillic acid, *p*-coumaric acid, *p*-hydroxybenzoic acid) from sterile soil (Cecil, Portsmouth and White Store) varied with soil type, horizon, time, and the type of phenolic acid added. When they allowed phenolic acids added to soil to equilibrate for 2 min before extraction, they noted a significant reduction in recovery of phenolic acids. Recovery declined with time up to 32 days. The decline was most rapid over the first 2 days. The presence of methoxy groups and acrylic side chains increased the irreversible sorption of these compounds in soil. Sorption in soil was generally in this order: ferulic acid > *p*-coumaric acid > vanillic acid ≥ *p*-hydroxybenzoic acid.

Recently Tharayil et al. (2006) using a Scitico silt loam soil (Typic Haplaquepts) and a batch equilibrium technique observed that *p*-coumaric, ferulic, *p*-hydroxybenzoic and vanillic acids all had strong site specific sorption, that soil organic matter was associated with preferential sorption, and that direct competition for sorption sites occurred even at low concentrations. For example, that 95% of sorbed vanillic acid was displaced into soil solution in the presence of ferulic acid, that *p*-hydroxybenzoic acid did not significantly affect the sorption affinity of the other three phenolic acids, and that ferulic acid showed low displacement by the other three phenolic acids. This differential sorption/desorption adds another level of complexity in that the concentration of an individual phenolic acid in the soil solution may be modified by the presence of other phenolic acids. In addition modification of sorption/desorption by competing phenolic acids will also influence the mobility of phenolic acids in soil. Furthermore Tharayil et al. (2008) demonstrated that decomposition of phenolic acids (i.e., residence time) in sandy loam soil (Typic Udorthent) and silt loam soil (Typic Udifluvent) occurred much more rapidly when single phenolic acids were added to these soils than when mixtures of phenolic acids (i.e., cosolutes) were added to these soils. Their conclusion: preferential sorption and degradation of phenolic acids in a mixture could protect and prolong the bioavailability of cosolutes and thus modify the overall toxicity of a mixture.

A word of caution about the procedures used to sterilize soils. Dalton et al. (1989a) compared the impact of autoclaving, methyl bromide and gamma-irradiation on Cecil, Portsmouth and White Store soils. They found that extractable amounts of Mn^{2+} from sterilized soils were two to eightfold greater than from non-sterilized soils. Recovery of ferulic acid from autoclaved-sterilized A-horizon soils was greater than recoveries from A-horizon sterilized soil by the other two techniques. Recoveries from methyl bromide treated soils and gamma-irradiated soils were similar. Amounts of ferulic acid extracted shortly after addition to soils, before microbial utilization was detectable, indicated that the amounts recovered from autoclaved-sterilized soils were more similar to non-sterilized soils than recoveries from methyl bromide and gamma-irradiated soils. These data suggest that sterilizing soils can lead to changes in soil physicochemistry (e.g., an increase in Mn, a multivalent cation) which can affect the sorption of phenolic acids to soil particles (e.g., reversible sorption by cation bridging) and that such changes in physicochemistry vary with soil type and sterilization technique (Salonius et al. 1967; Lopes and Wollum 1976; Williams-Linera and Ewel 1984; Dalton 1989; Dalton et al. 1989a). Others have used bioinhibitors such as $HgCl_2$ to inhibit microbial growth with little impact on other soil properties (Wolf et al. 1989; Tharayil et al. 2006). Aside from the safety issues the addition of Hg, a multivalent cation, may also modify sorption of phenolic acids. However no matter what technique is utilized, microbial activity must be curtailed or eliminated if data on soil sorption/desorption are to be meaningful.

2.4.3.4 Summary

When phenolic acids enter the soil environment they are reversibly and irreversibly sorbed to soil particles, polymerized, oxidized, reduced, leached, utilized by microbes, and taken up by roots. Rates for these various processes are highly variable and depend on soil type, biotic and physicochemical soil environment, types and mixtures of phenolic acids in or added to soils, and time, among others. To eliminate the effects of soil microbes, soils may be autoclaved. Concentrations of individual available phenolic acids in soils at a given point in time may be estimated by extracting soils with appropriate extractants and HPLC analysis. Based on our soils, we recommend water for estimating soil solution concentrations and neutral EDTA for soil solution and reversibly sorbed phenolic acid concentrations. However, the effectiveness of neutral EDTA in recovering available phenolic acids in all other soils should not be assumed. Reversibly sorbed phenolic acids increased or decreased as soil solution concentrations and multivalent cations increased or decreased, respectively.

2.4.4 Effects of Phenolic Acids on Bulk-Soil and Rhizosphere-Microbial Populations

Observations that phenolic acids were metabolized by soil microorganisms (Bonner and Galston 1944; Bonner 1946), that a variety of fungi and bacteria isolated from soils utilized phenolic acids as a carbon source (Henderson and Farmer 1955; Henderson 1956; Kunc 1971; Tack et al. 1972; Black and Dix 1976), and that enriching soils with phenolic acids stimulated microbial numbers, biomass and respiration (Henderson 1956; Kunc 1971; Turner and Rice 1975; Sparling et al. 1981) lead us to take a more in-depth look at the effects of phenolic acids on bulk-soil and rhizosphere microbial populations. The rhizosphere is generally small (on the order of mm) but varies considerably in distance depending on which rhizosphere characteristic is being measured (e.g., pH, water, nutrients, organic matter, and microorganisms). The rhizosphere is located between the rhizoplane of the root and the bulk soil. These terms are defined as follows:

a. rhizoplane – the surface of the root and the closely adhering/imbedded particles of soil;
b. rhizosphere – the narrow zone of soil adjacent to the root that is subject to the influence of living roots; and
c. bulk soil – soil beyond the rhizosphere that is not directly influenced by living roots.

For a more in depth description of these terms see Rovira (1991), Schmidt (1991), Bertin et al. (2003), Walker et al. (2003); Sylvia et al. (2004). In practical terms, microbial populations in the rhizosphere of cucumber seedlings were estimated as follows: root systems of each seedling were dipped briefly into 250 ml of sterile deionized water to dislodge large aggregates of soil material. The resulting root system with adhering soil material was then used to estimate rhizosphere microbial populations (see Section 2.2.11 for details; Shafer and Blum 1991).

2.4.4.1 Bulk Soil

Enriching (treating) Cecil or Portsmouth soils with > 0.125 μmol/g soil phenolic acid such as ferulic, *p*-coumaric, vanillic, and/or *p*-hydroxybenzoic, in the presence of adequate nutrition lead in a number of instances to an increase in microbial populations and the induction/selection of phenolic acid-utilizing microorganisms in the bulk soil (Blum and Shafer 1988; Blum et al. 1999a, b, 2000). Treatments with < 0.125 μmol/g soil lead to no detectable effects on microbial populations. In some instances, treatments with > 0.25 or 0.5 μmol/g soil lead to a decline in microbial populations. Observable effects varied with microbial type (e.g., bacteria, fungi, or actinomycetes), phenolic acid, phenolic acid concentration, soil type, and nutrient status of the soil. For example, maximum stimulation of bulk-soil bacterial populations for multiple treatments (given every other day) of 0.5 μmol/g soil of ferulic acid was 224% and 2784% in A- and B-horizon Portsmouth soil, respectively (Blum and Shafer 1988; Blum 2004). Fungi were stimulated 659% and 1762%, respectively. Actinomycetes were not significantly modified by ferulic acid enrichment. Maximum stimulation of bulk-soil bacterial populations by multiple treatments of 0.5 μmol/g soil of *p*-coumaric acid was 289% and 855% in A- and B-horizon Portsmouth soil, respectively (Blum and Shafer 1988; Blum 2004). Fungi were not stimulated in either soil and actinomycetes were not significantly modified in A-horizon soil but were inhibited 87% in B-horizon soil. As might be expected microbial responses to phenolic acid treatments of soils were not instantaneous. For example, maximum changes in Portsmouth soils treated with ferulic acid took days depending on the type of microbe being monitored (Fig. 2.19). However, since glucose was used as the carbon source in the selective media to estimate microbial populations, the observed changes included all organisms that could utilize glucose as a carbon source (Blum and Shafer 1988). How many of these organisms could actually utilize phenolic acids as a carbon source was not determined. In a subsequent experiment, the percent of bacteria that could utilize ferulic acid as a sole carbon source isolated from water, ferulic acid, nutrient solution, or ferulic acid-nutrient solution-treated B-horizon Portsmouth soil was 40%, 82%, 45%, or 92%, respectively.

Blum et al. (2000) isolated bacterial colonies that could utilize phenolic acid as a sole carbon source from Cecil A-horizon soils treated with individual phenolic acids, either *p*-coumaric acid or vanillic acid, and then tested these isolated bacterial colonies for their ability to utilize only *p*-coumaric acid, only vanillic acid or both phenolic acids. They found that the majority of isolates (> 72%) could utilize both phenolic acids while a much smaller fraction (< 28%) could only use the phenolic acid with which the soil had been treated. These results suggested that the majority of phenolic acid-utilizing bacteria in soils utilized a range of phenolic acids as a carbon source. Since soils contain a variety of phenolic acids, as well as other organic compounds (Flaig 1971; Paul and Clark 1989; Lavelle and Spain

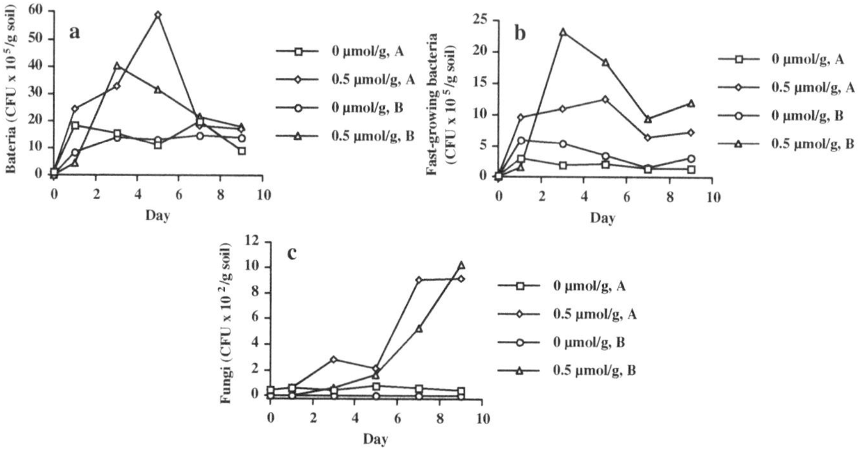

Fig. 2.19 Response of bacteria (**a**), fast-growing bacteria (**b**), and fungi (**c**) in Portsmouth A and B soils to 0 and 0.5 μmol/g soil ferulic acid applied every other day starting with day 1, where fast-growing bacteria represent colonies that were ≥ 1 mm in diameter after 6 days of incubation. For (**a**) $LSD_{0.05} = 2.9 \times 10^5$, for (**b**) $LSD_{0.05} = 2.88 \times 10^5$, and for (**c**) $LSD_{0.05} = 2.4 \times 10^2$. Figures reproduced from Blum and Shafer (1988)

2001; Strobel 2001), we subsequently determined changes in phenolic acid-utilizing bacterial populations after Cecil A-horizon soils were enriched with an equal-molar mixture composed of 7 phenolic acids plus or minus glucose (Blum et al. 2000). The addition of glucose did not modify the induction/selection and the microbial colony-forming units of phenolic acid-utilizing bacteria. This lead Blum et al. (2000) to conclude that the reduction of microbial utilization of phenolic acids observed in the presence of glucose by Pue et al. (1995) was due to preferential utilization of glucose over phenolic acids. Preferential utilization of one carbon source over another carbon source has also been observed by others in batch culture (Harder and Dijkhuizen 1982; Papanastasiou 1982) and in soils (Martin and Haider 1979; Sugi and Schimel 1993).

Proper mineral nutrition is also extremely important in determining the soil populations of bacteria that can utilize phenolic acids as a carbon source. Blum et al. (1999a) observed that when Cecil A-horizon soil (initially nutrient limited) was continuously supplied with 53 μg/ml (3.5 ml/h) of *p*-coumaric acid and a range of nutrient concentrations for 72 h by drip irrigation, the populations of phenolic acid-utilizing bacteria increased and then leveled off as nutrient concentration was increased. More specifically, phenolic acid-utilizing bacteria increased from 1.07 colony-forming units (CFU) $\times 10^5$/g soil to 3.55 CFU $\times 10^5$/g soil when the concentration of nutrient solution added to the columns was increased from 0 to 3%, a 282% increase. Increasing the nutrient concentration further from 3 to 50%, however, only raised the bacterial populations to 4.0 CFU $\times 10^5$/g soil, a 13% increase from the 3% nutrient treatment. In conjunction with this increase in phenolic acid-utilizing bacteria, the amount of *p*-coumaric acid collected from the base of the soil column decline in a linear manner with increasing bacterial populations (μg/ml *p*-coumaric acid recovered = 36.74 – 0.0000789 CFU, $r^2 = 0.89$, $p = 0.005$, where CFU equals colony-forming units of phenolic acid-utilizing bacteria).

Recently, Qu and Wang (2008) observed that, depending on phenolic acid concentration, microbial biomass carbon and basal respiration were either stimulated or inhibited by soil amendments of phenolic acids. In addition utilizing molecular techniques and cluster analysis they observed that such amendments caused significant shifts in fungal species diversity, more specifically changes in the number of dominant fungal types and fungal structural diversity.

2.4.4.2 Rhizosphere

Phenolic acid enrichment of soils containing roots (e.g., cucumber-microbe-soil systems) can also lead to an increase in rhizosphere microbial populations that can utilize phenolic acids as a sole carbon source (Shafer and Blum 1991; Blum et al. 2000; Staman et al. 2001). Shafer and Blum (1991) observed that the addition of up to 0.25 μmol/g of a phenolic acid to Portsmouth B-horizon soil did not detectably modify rhizosphere bacteria or fungi of cucumber seedlings but concentrations of 0.5 or 1 μmol/g soil did, suggesting that at the lower concentrations bulk-soil microorganisms and/or soil fixation prevented phenolic acids from reaching the rhizosphere or that other organic compounds dominated as the carbon source for the rhizosphere microorganisms. Blum et al. (2000) observed that the maximum stimulation of rhizosphere bacteria occurred when a 0.5 or 0.6 μmol/g of a 4- or 7-way equal-molar phenolic acid mixture, respectively, was supplied to Cecil A-horizon soil containing cucumber seedling roots (Fig. 2.20). Phenolic acid-utilizing bacteria increased by as much as 1200% (462% if data outliers are removed).

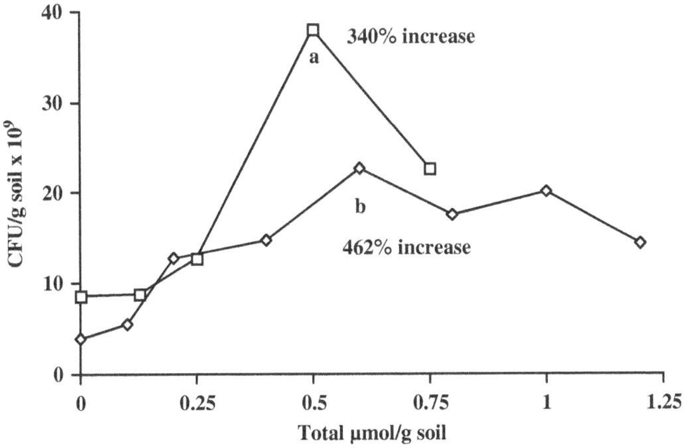

Fig. 2.20 The effects of multiple treatments of 7- (**a**) and 4- (**b**) equal-molar phenolic acid mixtures on cucumber seedling rhizosphere bacterial populations that can utilize phenolic acids as sole carbon sources, where CFU equals colony-forming units. Seedlings were grown in Cecil A soil. The 7-phenolic acid mixture was composed of caffeic, *p*-coumaric, ferulic, *p*-hydroxybenzoic, sinapic, syringic, and vanillic acids. The 4-phenolic acid mixture was composed of *p*-coumaric, ferulic, *p*-hydroxybenzoic, and vanillic acids. Figure based on data from Blum et al. (2000). Plenum Publishing Corporation, data used with permission of Springer Science and Business Media

2.4.4.3 Summary

Detectable population changes in laboratory systems of bulk-soil and rhizosphere phenolic-acid utilizing microorganisms to phenolic acid enrichment are a function of a variety of soil physicochemical and biotic factors including type of phenolic acid, phenolic acid enrichment concentrations, presence of other available organic compounds, nutrition, soil type, and initial microbial populations. Responses of microbes to phenolic acids or phenolic acid mixtures also varied with the type of microbe (e.g., bacteria, actinomycetes, or fungi).

2.4.5 Effects and Duration of Effects of Phenolic Acids on Seedlings in Soil Culture

Primary and short term effects resulting from phenolic acid treatments of seedlings are difficult to determine in soil culture because of the resistance of movement of phenolic acids in soil and the associated delay in distribution of phenolic acids throughout the soil system right after treatment (Schmidt and Ley 1999; Blum 2006). Secondary effects of phenolic acids in soil culture, which take a longer time to develop, do not have this problem. The inhibitory effects of phenolic acids on water utilization, evapotranspiration, leaf area expansion, and/or biomass varied with species (Einhellig et al. 1985; Waters and Blum 1987; Blum et al. 1993; Einhellig 2002; Shafer et al. 1998; Pue et al. 1995; Staman et al. 2001), phenolic acid (Blum et al. 1989), phenolic acid concentration (Blum et al. 1987, 1989), composition of phenolic acid mixtures (Blum et al. 1989), presence of other organic compounds (Blum et al. 1993; Pue et al. 1995), soil type (Blum et al. 1987), soil pH (Blum et al. 1989), and soil nutrition (Hall et al. 1983; Blum et al. 1993). Once phenolic acids were neutralized by irreversible sorption, microbial utilization, uptake by roots, etc., secondary processes recovered, the better the growth environment the faster the recovery (Blum and Gerig 2006). Effects of phenolic acids on seedling behavior described for soil systems were similar to what was observed in nutrient culture system (see Section 2.4.1). However, the magnitude of the response to a given concentration and the time for an observable response to phenolic acid treatments was lower and slower, respectively. These differences were largely due to the slower growth of seedlings in soil than in nutrient culture and soil associated processes (e.g., sorption, action of soil microbes) that impact the active concentrations of phenolic acids in soil solutions. For example:

a. Relative potency based on absolute growth rate of cucumber seedlings growing in Portsmouth soil were 1.00, 0.70, 0.67, 0.67, 0.59, 0.38, 0.35, and 0.13 for ferulic acid, sinapic acid, *p*-coumaric acid, vanillic acid, syringic acid, caffeic acid, *p*-hydroxybenzoic acid, and protocatechuic acid, respectively (Gerig and Blum 1991). Relative values, excluding ferulic acid, were on average 20–25% lower than in nutrient culture (Blum et al. 1985a), and

b. Absolute rates of leaf expansion of cucumber seedlings growing in Cecil A soil (pH 5.7) given a single 0.5 μmol/g soil *p*-coumaric acid (pH 5) treatment on the morning of day 6 were inhibited by 19% the next morning (day 7), reached a maximum inhibition of 34% by the morning of day 9, and were no longer inhibited by the morning of day 13 (Blum and Gerig 2006). The initial soil solution concentration for the 0.5 μmol/g soil *p*-coumaric acid treatment would be approximately 3 mM (based on 25 g water/150 g soil). That concentration in nutrient culture would be lethal to cucumber seedlings.

In nature soils do not contain single phenolic acids, but complex mixtures of phenolic acids, other organic compounds, and nutrients (Flaig 1971; Paul and Clark 1989; Lavelle and Spain 2001; Strobel 2001). It turns out that the action (i.e., inhibition) of individual phenolic acids on seedling processes were either partially antagonistic or additive in a mixture of phenolic acids (Blum et al. 1985a, 1989; Lyu et al. 1990; Gerig and Blum 1991; Lehman et al. 1994) and thus required concentrations of individual phenolic acids needed for a given percent inhibition declined as the number of phenolic acids increased in a mixture (Fig. 2.21; Gerig and Blum 1991; Blum 1996). In addition organic compounds, other than phenolic acids,

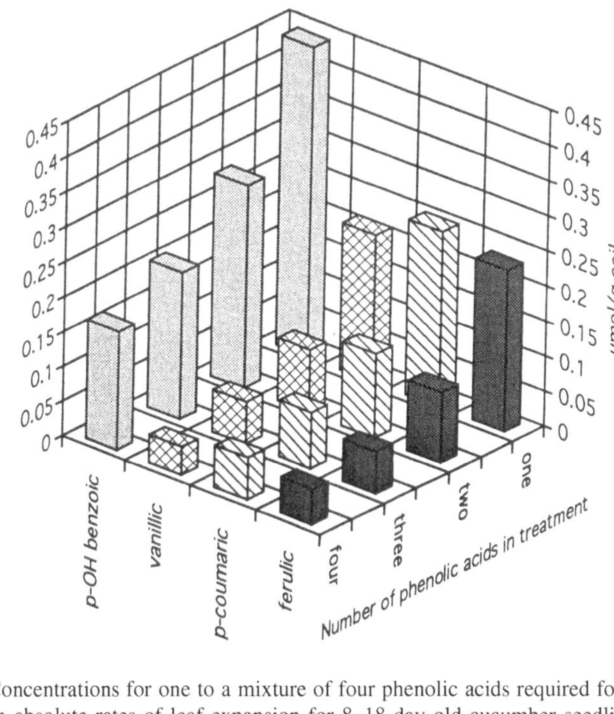

Fig. 2.21 Concentrations for one to a mixture of four phenolic acids required for a 30% inhibition of mean absolute rates of leaf expansion for 8–18 day old cucumber seedlings growing in Portsmouth B soil. Figure reproduced from Blum (1996). Figure used with permission of Society of Nematologists

and nutrients also modified the active concentrations of phenolic acids required for inhibition (Blum et al. 1993; Pue et al. 1995; Blum 1996). For example:

a. A 30% inhibition of absolute rates of leaf expansion of cucumber seedlings required 0.23 μmol/g soil of ferulic acid but required only 0.05 μmol/g soil in the presence of 0.06, 0.17, or 0.04 μmol/g soil *p*-coumaric acid, *p*-hydroxybenzoic acid, or vanillic acid, respectively (Gerig and Blum 1991; Blum 1996).
b. Methionine, glucose, and nitrate treatments influenced the inhibitory levels of *p*-coumaric acid on the biomass production of morningglory seedlings (Blum et al. 1993; Pue et al. 1995; Blum 1996). The addition of methionine and glucose reduced the concentration of *p*-coumaric acid required for a given percent inhibition (Fig. 2.22). For example, a 20% inhibition of morningglory biomass required 0.13 μmol/g soil of *p*-coumaric acid or 0.16 μmol/g soil of methionine but when both compounds were present, this inhibition was achieved with 0.063 μmol/g soil of *p*-coumaric acid and 0.054 μmol/g soil methionine (Blum et al. 1993; Blum 1996). In another study a 20% inhibition of morningglory biomass required 0.42 μmol/g soil *p*-coumaric acid in the absence of glucose and 0.33 μmol/g soil *p*-coumaric acid in the presence of 0.3 μmol/g soil glucose (Pue et al. 1995). Methionine at the concentrations used, were inhibitory while glucose concentration of < 4 μmol/g soil did not affect morningglory seedling biomass. Addition of nitrate, on the other hand, increased the *p*-coumaric acid concentration but reduced the methionine concentration required for a given percent inhibition (Blum et al. 1993; Blum 1996), and
c. The inhibition of red-root pigweed seedlings by chlorogenic acid was lost when Hoagland's solution was added in addition to chlorogenic acid (Hall et al. 1983).

Finally, similar to what had been observed in nutrient culture, effects of phenolic acids on cucumber seedlings were greater under acidic conditions than under neutral conditions, e.g., the dose required for a given percent inhibition of absolute and

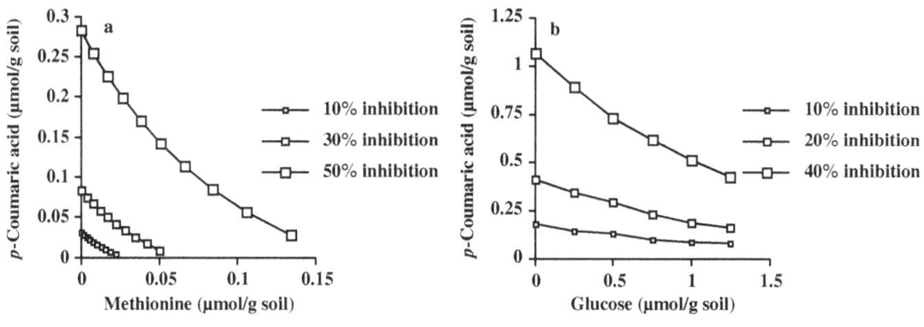

Fig. 2.22 Concentrations of *p*-coumaric acid and methionine (**a**), and *p*-coumaric acid and glucose (**b**) required to inhibit dry weight of morningglory seedlings growing in Portsmouth B and Cecil B soils, respectively, by 10–50%. Figures adapted/replicated from Blum et al. (1993) (**a**) and Pue et al. (1995) (**b**). Plenum Publishing Corporation, figures used with permission of Springer Science and Business Media

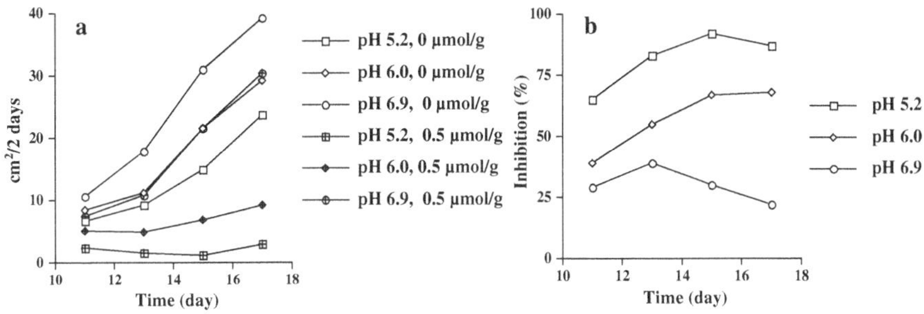

Fig. 2.23 Effects of ferulic acid on absolute growth rates (cm^2/2 days) of cucumber seedlings growing in Portsmouth A soil as modified by pH (**a**) and corresponding percent inhibition (**b**) calculated from data in (**a**). Figures based on data from Blum et al. (1989). Plenum Publishing Corporation, data used with permission of Springer Science and Business Media

relative rates of leaf expansion increased as the pH of the soil increased (Fig. 2.23; Blum et al. 1989). In addition the action (i.e., inhibition) of individual phenolic acids in a mixture of phenolic acids, particularly under acidic conditions, was more likely to be partially antagonistic. Under less acidic conditions or at very low concentrations their actions were much more likely to be additive (Blum et al. 1989; Gerig and Blum 1991).

In summary, observed effects of individual phenolic acids or phenolic acid mixtures were similar to what had been observed in nutrient culture but the response times and the magnitude of effects (see Section 2.4.8 for direct comparison) were slower and lower, respectively. Relative potencies of phenolic acids were lower when compared to nutrient culture. Increasing the number of phenolic acids in a mixture of phenolic acids reduced the concentrations of the individual phenolic acids required for a given percent inhibition. The presence of other readily available organic compounds (inhibitory or non-inhibitory) also reduced the concentration of phenolic acids required for a given percent inhibition. The addition of nitrate or nutrient solution reduced the inhibitory activity of phenolic acids. The inhibition of methionine, an amino acid, on the other hand was enhanced by the addition of nitrate. Finally phenolic acid effects were greater under acidic than under neutral conditions.

2.4.6 Relationships Between Phenolic Acid-Utilizing Microbes and Phenolic Acid Inhibition

That microorganisms can reduce the inhibitory effects of phenolic acids on seedlings has been observed by a number of researchers (Stowe and Osborn 1980; Sparling and Vaughan 1981; Sparling et al. 1981; Vaughan et al. 1983; Blum and Shafer 1988; Shafer and Blum 1991; Blum et al. 1987, 1999a, b, 2000; Staman et al. 2001; Inderjit 2005). I am, however, not aware of any study that has attempted to quantify how changes in bulk-soil bacteria might influence the inhibitory activity of phenolic acids. I am aware of only one study that has attempted to quantify how changes in rhizosphere microbial populations may influence the inhibitory activity of phenolic acids. Blum et al. (2000) observed that a 500% increase of phenolic acid-utilizing bacteria in the rhizosphere of cucumber seedlings growing in Cecil A-horizon soil enriched with an equal-molar mixture of 0.6 μmol/g composed of *p*-coumaric acid, ferulic acid, *p*-hydroxybenzoic acid, and vanillic acid resulted in a 5% reduction in the inhibition of absolute growth rates of its leaves (Fig. 2.24). This is much less than would be expected by some (Schmidt 1988; Schmidt and Ley 1999), but not by others (Williamson and Weidenhamer 1990).

Utilization of phenolic acids by microbes can be substantial as long as conditions are appropriate, e.g., adequate nutrition, moisture, pH, and temperature. However, a variety of other factors can also influence microbial utilization of phenolic acids. For example, phenolic acid utilization by microbes was reduced when more readily available carbon sources were present in the soil such as glucose and phenylalanine (Fig. 2.24; Blum et al. 1993; Pue et al. 1995) or other sources of available carbon, i.e., roots (Blum et al. 1999a). The presence of methionine, however, did not influence the rate of phenolic acid utilization by soil microbes (Pue et al. 1995). Differential utilization of carbon sources by soil microbes is fairly common (Martin and Haider 1979; Harder and Dijkhuizen 1982; Papanastasiou 1982; Sugi and Schimel 1993). Thus the relationship between phenolic acid-utilizing microbes and their carbon environment is complex. There is yet another aspect that adds to this complexity. Initial microbial breakdown products of phenolic acids are frequently other phenolic acids. For example, ferulic acid is converted to caffeic acid or vanillic acid and these are converted to protocatechuic acid. Next the ring structure of the

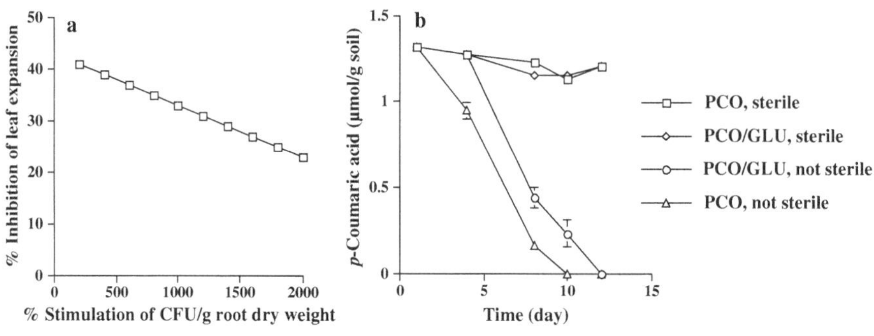

Fig. 2.24 Relationships (**a**) between percent stimulation of rhizosphere bacteria that can utilize phenolic acids as sole carbon sources and percent inhibition of absolute rates of leaf expansion of cucumber seedlings growing in Cecil A soil treated with a 0.6 μmol/g soil 4-equal-molar phenolic acid mixture (**a**; $r^2 = 0.50$), where CFU equals colony-forming units and the 4-phenolic acid mixture was composed of *p*-coumaric acid, ferulic acid, *p*-hydroxybenzoic acid, and vanillic acid. The recoveries (**b**) of "free" and reversibly sorbed *p*-coumaric acid (PCO) from sterile or non-sterile Cecil B soil in the presence or absence of glucose (GLU). The absence of standard error bars for (**b**) indicates that the error bars are smaller than the *symbols* representing the mean. **a** was based on a regression from Blum et al. (2000) and **b** was reproduced from Pue et al. (1995). Plenum Publishing Corporation, regression and figure used with permission of Springer Science and Business Media

protocatechuic acid ring is broken to form *ß*-carboxy-*cis, cis*-muconic acid which is then converted to ß-oxoadipic acid which in turn is broken down to acetic acid and succinic acid and these are ultimately broken down to CO_2 and water (Evans 1963; Dagley 1971; Martin and Haider 1976). Each of these phenolic acids has different microbial utilization rates, different sorption rates in the soil, and different inhibitory effects on seedlings (Blum 1998, 2004, 2006; Pue et al. 1995). These differences in behavior for individual phenolic acids and their breakdown products can substantially influence the magnitude and duration of the observed seedling inhibition. Thus the inhibitory activity observed for a given phenolic acid represents the action of a complex set of interacting compounds, not just the original phenolic acid (Blum et al. 1993, 1994; Blum 1996).

Finally, Blum et al. (2000) also observed a rather surprising relationship between inhibition and phenolic acid utilizing rhizosphere microbes. They observed that there was an inverse-linear relationship between phenolic acid-utilizing bacteria in the rhizosphere and the absolute growth rate of cucumber leaf expansion up to 0.6 or 0.7 μmol/g soil for a 4-phenolic acid mixture (Fig. 2.25), or restated, a linear relationship between increasing microbial populations and the inhibition of leaf expansion. The expectation would be that as the microbial populations that can utilize phenolic acids as a carbon source increase in the rhizosphere that the inhibition of leaf expansion by phenolic acids would decline (Blum et al. 2000). The most likely explanation in this case is that the effects of phenolic acids on seedlings and microbes in the rhizosphere were independent. This independence could occur whenever the concentrations of phenolic acids in the rhizosphere or the rates at which phenolic acids reach the rhizosphere are in excess of what can be utilized by microbes in the rhizosphere. In another study using ground wheat or sunflower tissues as a source of phenolic acids Staman et al. (2001) also found this inverse relationship between cucumber leaf expansion and phenolic acid-utilizing rhizosphere bacteria. However, simultaneous inhibition of seedling growth and

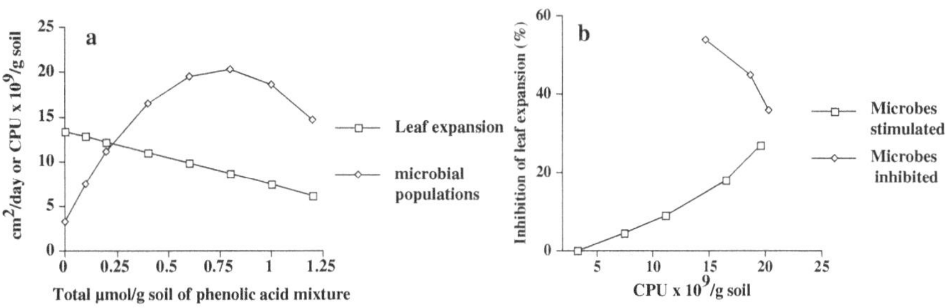

Fig. 2.25 Effects of total phenolic acid composed of a 4-equal-molar mixture of *p*-coumaric acid, ferulic acid, *p*-hydroxybenzoic acid, and vanillic acid on absolute rates of leaf expansion (cm^2/day; $r^2 = 0.44$) of 12 day-old cucumber seedlings and microbial populations (CFU/g soil; $r^2 = 0.49$) that can utilize phenolic acids as a sole carbon source in Cecil A soil (**a**). Relationships between phenolic acid-utilizing microbes (CFU, colony-forming units) and percent inhibition of absolute rates of leaf expansion for cucumber seedlings are presented in **b**. Values for (**b**) were calculated from values in (**a**). Figures based on regressions from Blum et al. (2000). Plenum Publishing Corporation, regressions used with permission of Springer Science and Business Media

stimulation of microbes within the rhizosphere that utilize phenolic acids as a sole carbon source was inconsistently expressed over time for these tissue additions (for additional details see Section 3.4.7).

In summary phenolic acids are readily utilized by soil microbes. However rates of utilization are highly variable depending on the biotic and physicochemical soil environment including the presence or absence of other readily available organic compounds. In the presence of some, but not all, readily available organic compounds, the microbial utilization of phenolic acids was reduced. Utilization of phenolic acids by microbes is rather complex in that initial breakdown products are also phenolic acids and, therefore, inhibition of seedlings observed is a function of the original phenolic acid and its phenolic acid breakdown products as well as other subsequent breakdown products. In general, however, as populations of phenolic acid-utilizing microbes increase the effect of a given phenolic acid concentration treatment declines, but given the right conditions effects of phenolic acids on seedlings and soil microbes can be independent. In the later case simultaneous stimulation of phenolic acid utilizing microbes and inhibition of seedling processes can occur.

2.4.7 Effects of Seedling-Microbe-Soil Systems on the Available Concentrations of Phenolic Acids in Soil Solutions

Concentrations of phenolic acids in natural soils appear to change slowly, particularly when a well developed litter layer is present (Turner and Rice 1975; Lodhi 1978; Whitehead et al. 1979; Blum et al. 1991). However, concentrations of available phenolic acids applied to cucumber-microbe-soil systems declined very rapidly (Klein and Blum 1990b; Blum et al. 1987, 1999a, b; Blum and Gerig 2006). In most instances depending on treatment concentration, nutrient level, and the age of the system, available phenolic acids were extremely small or could not be recovered by neutral EDTA extraction 24–48 h after the initial phenolic acid treatment. To approximate phenolic acid concentrations in natural soils, therefore, required continuous additions (Blum et al. 1999a) or multiple additions at given time intervals of phenolic acids (Blum et al. 1987; Blum and Gerig 2006). For continuous addition, i.e., the flow-through system, phenolic acid and nutrients were supplied by drip irrigation (2–3.5 ml/h; Blum et al. 1999a). For multiple additions water was added to a constant level every 24 h after nutrient or phenolic acid solutions had been added. Nutrient solutions and phenolic acid solutions were added on alternate days. No matter the method of phenolic acid application, phenolic acids applied to soils were rapidly sorbed by soil particles, utilized by microbes and/or taken up by roots. For the flow-through system some losses due to leaching also occurred over the first 36–48 h (Blum et al. 1999a). The bottom line, in spite of these techniques, available phenolic acid concentrations in the cucumber-microbe-soil systems varied over time. For multiple additions the highest available concentrations of phenolic acids in soil were observed right after treatment and the lowest just before the next treatment (Blum et al. 1987; Blum and Shafer 1988; Pue et al. 1995; Blum and Gerig 2006).

For the flow-through system phenolic acid (*p*-coumaric acid) recovered from the base of the seedling-microbe-soil columns (i.e., leached) declined over time to zero recovery. The lower the phenolic acid concentration applied the faster the decline to zero (Fig. 2.26; Blum et al. 1999a). Soil extractions with water and citrate at this point did not recover any phenolic acids, i.e., phenolic acids were being taken up by roots, utilized by microorganisms, and/or irreversibly sorbed by soil particles as rapidly as they were being supplied to the flow-through system. For microbe-soil systems (no seedlings), phenolic acid recovered from the columns initially increased (first 24 h), but then rapidly declined over time until they reached a steady state (Fig. 2.26). The rate of decline and steady-state concentration in this case varied with the nutrient level supplied to the soil-microbe system. As long as nutrients were limiting, the higher the nutrient level applied the more rapid the decline and the lower the steady-state concentration (Blum et al. 1999a). Losses ranged from 36 to 87% once the system reached a steady state. For autoclaved (sterile) soil, recovery initially increased until the system was saturated (roughly after 36 h) and then recovery remained constant (Fig. 2.26). Reversibly sorbed *p*-coumaric acid at steady state for sterile Cecil A soil increased with increasing nutrient levels supplied and ranged from 25 to 42%. Approximately 12% of the phenolic acid was irreversibly sorbed once soils were saturated. This suggested that the primary losses observed for the cucumber seedling-microbial-soil systems were due to microbial utilization and/or root uptake. We suspect that of the two the microbial utilization was the more important particularly since soil conditions were suitable for microbial activity.

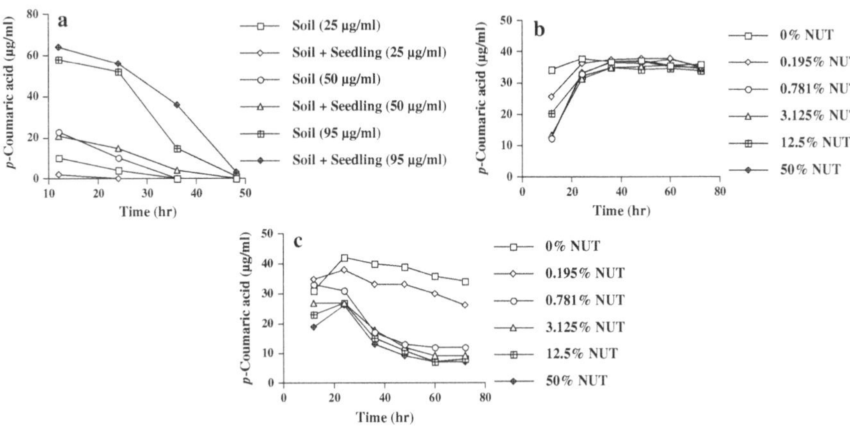

Fig. 2.26 Recoveries of *p*-coumaric acid from the bottom of Cecil A soil columns in the presence of cucumber seedlings and microbes (**a**), in the absence of microbes and seedlings (**b**), and in the presence of microbes but absence of seedlings (**c**). For (**a**), approximately 25, 50 or 95 μg/ml of *p*-coumaric acid in 25% Hoagland's nutrient solution was applied to the columns at a rate of 2–3.5 ml/h. For (**b**) and (**c**), 41 and 54 μg/ml, respectively, of *p*-coumaric acid in different nutrient solution concentrations (0–50%) was applied to columns at the same rate as in (**a**). Figures reproduced from Blum et al. (1999a). Cádiz Univ Press, Puerto Real. Figures used with permission of Servicio de Publicaciones Universidad De Cádiz

There is an additional aspect regarding soil solution concentrations and evapotranspiration that should be highlighted particularly in closed soil systems such as our cup system (Blum and Gerig 2006). Essentially it is this. As phenolic acid concentrations applied to seedling-microbe-soil-systems increased, the greater the reduction in evapotranspiration (primarily transpiration) and the slower the decline of the water content of the soil. The average soil water content over any 24-h watering period was, therefore, related to the concentration of phenolic acid added. Thus over any given 24-h watering period, the soil solution concentration of a phenolic acid expressed in mM phenolic acid varied over time (initially decreasing fairly rapidly and then slowly increasing as water was depleted; Fig. 2.27). These mM concentrations were different than when the concentrations of phenolic acid were expressed as μmol of phenolic acid/g soil (concentration in this instance decreased rapidly at first and then began to slow and level off towards the end of the 24-h period; Fig. 2.27). If we add to that that the magnitude of the effects of phenolic acids or phenolic acid mixtures appear to be directly proportional to the root system in contact with phenolic acids (Klein and Blum 1990a; Lyu and Blum 1990; Lehman et al. 1994; Lehman and Blum 1999b), we are left to conclude that average mM soil solution concentrations are more meaningful than average μmol/g soil concentrations when attempting to relate soil concentrations of phenolic acids to seedling or microbial behavior. But even this approach is not entirely satisfactory since:

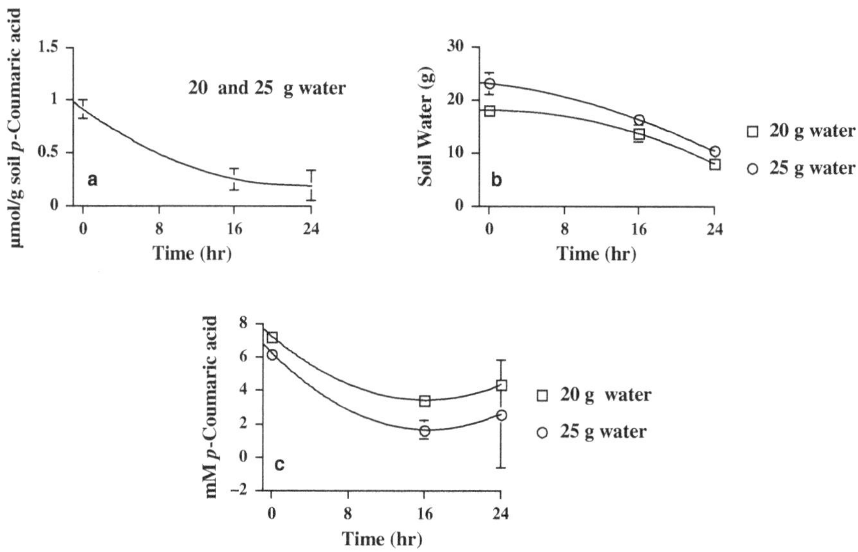

Fig. 2.27 The changes in μmol/g soil *p*-coumaric acid (**a**), soil water (g/150 g soil) (**b**), and mM *p*-coumaric acid (**c**) for cup systems with 12–13 day-old cucumber seedlings and Cecil A soil. Systems were treated with 1 μmol/g soil *p*-coumaric acid and 20 or 25 g water/150 g soil. Absence of error bars indicates that error bars are smaller than the *symbols* representing the mean. Figures reproduced from Blum and Gerig (2006). Figures used with permission of Springer Science and Business Media

(a). higher soil moisture resulting from the inhibition of evapotranspiration lowered soil solution concentrations of phenolic acids by 14–40% but did not significantly influence the inhibitory effects of seedlings within the time frame of a 13-day experiment (Blum and Gerig 2006), and
(b). in the open-flow system (Blum et al. 1999a) inhibition of cucumber seedlings occurred even when no phenolic acids were detected in the soil of the flow-through system (i.e., input was balanced by loss).

In summary because of adequate nutrition and soil moisture the loss of available phenolic acids in the laboratory soil systems was much more rapid than in field soils. In order to maintain some consistency in phenolic acid concentrations over time, phenolic acids were added on alternate days (closed cup system) or supplied continuously (flow-through system). For the cup system soil solution mM concentrations over time were not only determined by the concentration added and soil processes (e.g., sorption, microbial utilization, and root uptake), but also by phenolic acid modified soil moisture, i.e., with increasing concentrations of *p*-coumaric acid average soil moisture increased and average mM of *p*-coumaric acid decreased. In addition the sterile flow-through system demonstrated that soil can be saturated by phenolic acid and that irreversible soil fixation at those times is negligible. Losses observed in the cucumber seedling-microbial-soil systems were largely due to microbial utilization and/or root uptake. Of the two, microbial utilization appeared to be the more important.

2.4.8 Comparison of the Effects of Phenolic Acids on Seedlings in Nutrient and Soil Culture

What becomes evident almost immediately when comparing cucumber seedling behavior in nutrient culture and in soil culture is that response times and the level of response to phenolic acid inputs or losses are much faster and consistently greater in nutrient culture than in soil culture. For example:

a. The time required for initial detection of inhibition by a single *p*-coumaric acid treatment to soil was 1 day for leaf area expansion and up to 3 days for evapotranspiration (Blum and Gerig 2006). Since *p*-coumaric acid did not affect evaporation from soil, inhibition of evapotranspiration actually represents inhibition of transpiration. Maximum expression of inhibition required 1 and 6 additional days, respectively. Recovery occurred shortly after maximum expression, but full recovery required an additional 2–5 days. In nutrient culture maximum expression of inhibition of leaf area expansion and transpiration and subsequent recovery all occurred within 1–4 days (Blum and Dalton 1985; Blum et al. 1985a; Blum and Rebbeck 1989).
b. A 20% inhibition of absolute rates of leaf expansion for 12 day old cucumber seedlings treated on day 6, 8 and 10 with *p*-coumaric acid required 0.14 mM (2.53 mg added/treatment; 110 ml) of *p*-coumaric acid in nutrient culture (Blum and Gerig 2005) and 0.55 μmol/g soil (13.55 mg added/treatment; 150 g soil,

20 ml water) in soil culture (Blum and Gerig 2006). The 0.55 μmol/g soil would be equivalent to 4.12 mM at time 0 in the soil solution. The required mM of *p*-coumaric acid added to the soil system was, therefore, 29 times greater for the soil system than for the nutrient system, and

c. The relative potencies of phenolic acids in soil culture were also generally lower from those observed in nutrient culture. Relative potency based on absolute growth rate of cucumber seedlings growing in Portsmouth soil were 1.00, 0.70, 0.67, 0.67, 0.59, 0.38, 0.35, and 0.13 for ferulic acid, sinapic acid, *p*-coumaric acid, vanillic acid, syringic acid, caffeic acid, *p*-hydroxybenzoic acid, and protocatechuic acid, respectively (Gerig and Blum 1991). Relative values, excluding ferulic acid, on average were about 20–25% lower than in nutrient culture with a range of 0–63% (Blum et al. 1985a).

The higher concentrations of phenolic acids required for a given percent inhibition between the two systems stem from the fact that nutrient cultures have a much more consistent environment than soil culture systems in that water, nutrients, and phenolic acids are evenly distributed in the treatment container and thus are readily available to interact with root surfaces. Soil systems, on the other hand, are much more complex heterogeneous environments in which roots must compete with a variety of soil sinks (e.g., clays, organic matter, and microbes) for water, nutrients, and phenolic acids. There is also mechanical resistance to the movement of water, nutrients, and phenolic acids and the growth of roots in soils. The slower development of inhibition after treatment and the slower recovery after phenolic acid depletion in soil systems is very likely related to the slower growth of seedlings in soil culture.

In summary inhibition of phenolic acid treatments took longer to develop and took higher concentrations in soil culture than in nutrient culture. Recovery after phenolic acid depletion in the root zone was also slower in soil culture than in nutrient culture.

2.4.9 Effects of Phenolic Acids at Various Life Stages

Cucumber plants do not lend themselves very well to life cycle studies in the laboratory environment because of their growth habit and size at maturity. We thus chose to use *Phaseolus vulgaris* L. "BBL-290", a bush snap bean with a more appropriate growth habit and a short life cycle to determine how the various stages of a life cycle might respond to the treatment of phenolic acids. BBL-290 produces flowers at approximately 24 days, pods at 30 days, and dry seeds at 70 days after planting in a growth chamber (Blum and Heck 1980). Unfortunately bush snap bean seedlings are much less sensitive to phenolic acids than cucumber seedlings (Waters and Blum 1987; Holappa and Blum 1991). However, based on single and multiple ferulic acid exposures it would appear that the seedling and flowering stages are the most sensitive stages in the life cycle of bush snap beans. This it turns out is the only study that has attempted to address this issue. A quick review of the literature suggests that plant–plant allelopathic interaction studies have focused primarily on the earlier stages of plant growth cycle, i.e., germination and seedling development. This is not entirely surprising since germination and early seedling establishment and development are particular "weak links" or susceptible stages to stress in plant life cycles. Fully developed plants have a greater buffering capacity for stress and have larger deeper root systems reducing the likelihood that a large portion of their root system will come in contact with inhibitory phenolic acid concentrations.

During reproductive stages (i.e., flowering, fruit and seed development) vegetative growth is greatly reduced or has stopped and water, nutrients, and energy resources are now predominately directed towards the reproductive structures. However, when reproductive structures compete for these resources, which they frequently do, even small modification of a plant's source-sink relationships (e.g., water, nutrients, and/or energy) by phenolic acids can have a substantial impact.

In summary based on the bush snap bean life cycle, early seedling development and reproduction, particulary at flowering, are most sensitive stages to phenolic acids.

2.5 Summary of Major Points for Model Systems

2.5.1 Seedlings

a. Primary effects (e.g., root membranes perturbations) of simple active phenolic acids, such as cinnamic and benzoic acid derivatives, were localized to their initial site of contact (e.g., root membrane) and were thus not systemic in nature. Effects were directly related to the percent of the root system in contact with phenolic acids and poorly related to phenolic acid uptake by roots. Active phenolic acids consist of those phenolic acids that actually interact with root membranes.
b. Membrane perturbations lead to a reduction in hydraulic conductivity (i.e., reduced water uptake, decreased water potential and turgor pressure) and net nutrient uptake, increased efflux of ions, and a host of other secondary effects (e.g., reduced photosynthesis and growth).
c. Response times for effects of phenolic acids varied for processes being observed. For example, membrane perturbations occurred in seconds, reductions in rates of leaf area expansion occurred within 24–48 h, and reductions in biomass occurred within 5–7 days. Response times also varied with growth medium, e.g., response times were faster in nutrient culture than in soil culture.
d. The magnitude of effects by active phenolic acids were a function of seedling sensitivity, "health", and state of acclimation, proportion of root system in contact with phenolic acids, stage of life cycle, type of phenolic acid, numbers of phenolic acids, concentration and duration of active phenolic acids interacting with roots, stresses before or during phenolic acid treatments, types of other organic and inorganic substances present, growth medium, pH and microbial population of the bulk soil, rhizosphere, and rhizoplane.

e. The inhibitory activity of individual phenolic acids varied with seedling species (i.e., taxon). However, cinnamic acid derivatives were generally more inhibitory than their equivalent benzoic acid derivatives.
f. Simple phenolic acids were taken up by roots and were translocated throughout the seedling. However, active phenolic acids were short lived in plant tissues since the actions of phenolic acids were rapidly neutralized by sorption to cell walls, incorporation into lignin, glucosylation, metabolism, and/or sequestration into vacuoles. Thus, the majority of the phenolic acids taken up by roots remained within the root tissues.
g. Uptake of phenolic acids by seedlings was highly variable and varied with species, seedling age, phenolic acid concentration, type and number of phenolic acids present, pH of nutrient or soil solution surrounding roots, pH of rhizosphere and rhizoplane, time of day, time and number of treatments, type of uptake (apoplastic vs. symplastic), level of aeration, and prior water, nutrient and/or phenolic acid stresses.
h. Seedlings modify active phenolic acid concentrations surrounding their roots by sorption of phenolic acids to cell walls, uptake of phenolic acids by roots, and changing the pH of the rhizoplane and rhizosphere, a type of feedback regulation by the seedlings.
i. Primary and many secondary inhibitory effects of simple phenolic acids were readily reversed once root contact with active phenolic acids was reduced or eliminated (e.g., by seedling root uptake, microbial utilization, leaching, and/or removal). Examples of processes that were readily reversed were transpiration, net nutrient uptake, photosynthesis, and leaf expansion. Magnitude of recovery and the decline of active concentrations of phenolic acids were closely linked. Recovery was faster for seedlings growing in an adequate water/nutrient environment than for seedlings growing in a limited water/nutrient environment, and was faster in nutrient culture than in soil culture systems. Depending on the seedling process being observed recovery times ranged from seconds (e.g., recovery of membranes) to days (e.g., recovery of seedling biomass). However, anatomical and morphological characteristics once established did not recover, and
j. Based on bush snap bean life cycle, early seedling development and reproduction, particularly at flowering, were the most sensitive stages to phenolic acids.

2.5.2 *Microbes*

a. Seedling rhizoplane and rhizosphere and bulk soils contain microbes that can utilize simple phenolic acids as a carbon source.
b. Enrichment of soils (or nutrient cultures) with phenolic acids can result in an induction/selection of microbes in the bulk soil, in the rhizosphere, and on the rhizoplane that can utilize phenolic acids as a sole carbon source.

c. Induction/selection of microbes varied with microbial type (e.g., bacteria, fungi, and actinomycetes), phenolic acid, phenolic acid concentration, medium (e.g., type of soil), and nutrient status of the medium.
d. Minimum and maximum enrichment concentrations were observed for the induction/selection process. These minimum and maximum concentrations varied with microbial type, phenolic acid, medium, and nutrient status of the medium.
e. The majority of phenolic acid-utilizing bacteria (and we suspect other types of microbes as well) isolated from soil could utilize a range of simple phenolic acids as a carbon source, i.e., were generalists.
f. Utilization of phenolic acids by soil microorganisms was substantial as long as environmental conditions were appropriate. However, a variety of other factors also influenced microbial utilization of phenolic acids. For example, utilization of phenolic acids was reduced when other more readily available carbon sources were present and increased with aeration. In the absence of aeration there was an accumulation of phenolic acid breakdown products. Seedling inhibition of resulting phenolic acid breakdown products were equal to or less than the original phenolic acid but in many instances they were less inhibitory, and
g. As phenolic acid-utilizing bacteria in the rhizosphere increased, inhibition of absolute rates of leaf expansion of cucumber seedlings declined, but given the right conditions effects of phenolic acids on seedlings and microbes in the rhizosphere can be independent. In the later case simultaneous stimulation of phenolic acid utilizing microbes and inhibition of seedling processes can occur.

2.5.3 Phenolic Acids

a. Phenolic acids in nutrient culture are in solution, i.e., a "free" state. They may be utilized by microbes, taken up by roots, oxidized and/or reduced. Concentrations of phenolic acids in nutrient culture were determined by HPLC analysis of nutrient solution subsamples. Depletion rates of phenolic acids in nutrient culture varied with the biotic components present (e.g., seedling species, types of microbes present), number, type, and concentration of phenolic acids present, and the physicochemical environment (e.g., pH, and aeration).
b. Phenolic acids in the soil solutions, i.e. "free" state, are reversibly and irreversibly sorbed to soil organic and inorganic soil particles, utilized by soil microbes, taken up by roots, move vertically and horizontally by capillary movement or "transpirational pull" of the soil solution, leached by gravitational water, oxidized, reduced, and/or polymerized. Rates for each of these processes varied or appeared to vary with soil type, phenolic acid, phenolic acid concentration, composition of a phenolic acid mixture, and the biotic (e.g., microbes) and physicochemical environment.
c. Concentrations and fractions of simple phenolic acids recovered from soils varied with the extraction procedure utilized. Depending on procedure utilized,

fractions recovered ranged from "free" phenolic acids in the soil solution to irreversibly sorbed phenolic acids on/in organic and inorganic soil particles.

d. Reasonable estimates of concentrations of individual "free", reversibly sorbed, and available ("free" and reversibly sorbed) phenolic acids in our soils were obtained by water and neutral EDTA extractions and quantification by HPLC analysis. Recovery of phenolic acids by water and neutral EDTA varied with phenolic acid, phenolic acid concentration, soil type, soil horizon, soil nutrients and organic matter, size of phenolic acid-utilizing microbial populations, extraction time, and extractant-soil slurry pH.
e. As "free" phenolic acids were depleted from soil solution, reversibly sorbed forms were released from organic and/or inorganic soil particles back into the soil solution. As "free" phenolic acids in soil solutions increased, more phenolic acids were reversibly sorbed to organic and/or inorganic soil particles. Reversible sorption also increased or theoretically decreased as multivalent cations increased or decreased within the soil, respectively.
f. Available vs. active fractions: The available fraction was found to be composed of "free" phenolic acids in solution (soil solution or nutrient solution) and reversibly sorbed phenolic acids on soil particles. The available fraction constitutes the phenolic acids in the soil or nutrient solution that potentially can interacts with the roots and microorganisms. The active fraction constitutes that portion of the available phenolic acids in the soil or nutrient solution that actually interacts with the roots and microorganisms.
g. Estimates of concentrations of available phenolic acids in the bulk soil may only be meaningful if they are related or correlated to concentrations in the rhizosphere/rhizoplane. The rhizosphere may in fact be a very strong "barrier" between phenolic acids in the bulk soil and the root surface (rhizoplane). Because of the nature of phenolic acids in nutrient solution, available concentrations in the nutrient solution and next to the rhizoplane are essentially identical. Thus the whole nutrient solution constitutes the rhizosphere. The rhizosphere in soil systems is generally represented by a zone limited to mm around the roots and is thus dissimilar from the bulk soil.
h. pH and phenolic acid effects: One hypothesis states that the active concentrations of phenolic acids decline as acidity of the nutrient solutions and soil solutions surrounding roots, rhizosphere and/or rhizoplane decline (become less acidic). The reason – under neutral or basic conditions simple phenolic acids become negatively charged (i.e., ionized) and will not interact with or be taken up by roots and microbes, that also tend to be negatively charged. Maximum inhibition would require an acidic environment. A second hypothesis states that decreasing pH protonates the functional –OH groups on cell walls to form positively charged $-OH_2^+$ groups which then bind the negatively charged phenolic acids. The increasing electrostatic binding of phenolic acids to cell walls with decreasing pH then directly or indirectly leads to an increase in phytotoxicity by modifying the composition of the solution reaching the cell membranes and/or by modifying cell wall chemistry.

i. Concentrations of phenolic acids are frequently expressed as μg, μM, or mM in nutrient culture and μg/g soil or μmol/g soil in soil culture. Since molecular weights vary with each phenolic acid, the use of μM, mM and μmol/g soil is more meaningful if comparisons are to be made among phenolic acids. However for soils, water content varies over time and thus concentrations contacting roots vary accordingly. For this reason the use of μmol/g should be minimized and the use of μM or mM should be encouraged wherever possible. Since phenolic acids can inhibit water uptake and transpiration by seedlings and thus slow the rate of decline for soil water, the use of μM or mM becomes even more important.

j. Roots must interact with phenolic acids before an effect on seedlings can occur. Thus estimates of concentrations of active phenolic acids in the rhizosphere/rhizoplane are required. Unfortunately we presently do not have a means of quantifying the active phenolic acid fraction, thus the reliance on the available phenolic acid fraction.

k. Inhibitory effects and relative potency of individual phenolic acids were lower in soil culture than in nutrient culture. Time required for inhibition and subsequent recovery were shorter in nutrient culture than in soil culture. These differences between nutrient culture and soil culture were largely due to the action of soil sinks (e.g., clays, organic matter, and microbes), lower phenolic acid mobility in soil, and the overall slower growth rate of seedlings in soil.

l. Mixtures of active phenolic acids in nutrient solutions and soil solutions were inhibitory even when concentrations of individual phenolic acids were well below their inhibitory levels but the total concentrations of all active phenolic acids in the mixture were at inhibitory levels. Effects of individual phenolic acids in mixtures were found to be either additive (at lower concentrations or higher pH) or partially antagonistic (at higher concentrations or lower pH).

m. The available/active concentrations of phenolic acids surrounding roots were a function of input (i.e., enrichment and subsequent microbial conversion and synthesis), losses (e.g., sorption by soil particles, microbial conversion and utilization, polymerization, oxidation-reduction, root uptake, and leaching below root zone), and state of protonation. Initial input (enrichment for our model systems) was predetermined but losses (i.e., depletion) were highly variable depending on type and number of phenolic acids present, number and time of phenolic acid treatments, initial phenolic acid concentrations, presence of other available organic compounds, seedling species, age of seedling, size of phenolic acid-utilizing microbe populations, nutrition, and soil environments to name a few.

n. The presence of other organic and inorganic compounds, depending on concentrations and types present, modified the inhibitory effects of simple phenolic acids. For example, a 10% inhibition of morningglory biomass required 0.046 μmol (7.5 μg) *p*-coumaric acid/g soil while in the presence of 0.025 μmol (3.68 μg) methionine/g soil the *p*-coumaric acid concentration required was only 0.023 μmol/g soil (3.75 μg/g soil). Similar response trends were obtained for *p*-coumaric acid and glucose. Adding nitrate to soil lowered the concentration of methionine required but increased the concentration of *p*-coumaric acid required for a 10% inhibition of morningglory seedling biomass.

o. In the absence of soil microbes and seedlings, enrichment of soils with phenolic acids by continuous addition (flow-through system) eventually saturated the soil with phenolic acids. In the Cecil A soil flow-through system, depending on nutrition, reversible sorption ranged from 25 to 42% and irreversible sorption was approximately 12% for Cecil A soil (1 soil:2 sand mixture) at steady state, and

p. Effects of phenolic acid on cucumber seedlings in the flow-through system were observed even when no phenolic acids could be extracted by water or citrate from the soil in which the seedlings were growing.

2.6 Relevance of Model Systems to Field Studies

The focus of the research just described utilizing the cucumber-nutrient culture and the cucumber-microbe-soil model systems was to identify and characterize what happens to phenolic acids once they enter the soil environment and how in turn this may influence the behavior and processes of phenolic acid-sensitive seedlings. The research was thus hypothetical, theoretical, and process oriented and not designed to answer specific questions related to actual field observations. When choosing any model system for study one hopes and assumes that the model system chosen is not unique but that the relationships and processes observed for the system chosen are representative of a broader array of other systems and hopefully relevant to field systems.

How representative a model system is can be determined to some extent by testing other bioassay species and soils in the model system chosen and by comparing the biological and physicochemical characteristics and processes of model and field systems. The following considerations are thus worth noting:

a. In a number of experiments bush snap bean, corn, morningglory, pigweed, or tomato seedlings were substituted for cucumber seedlings. Species substitutions in the model system resulted in consistent behavior of sensitive species for the processes observed (i.e., the model cucumber nutrient culture and cucumber-microbe-soils systems were not unique in their behavior).

b. The nutrient culture and soil model systems used were designed to minimize environmental stress and thus maximize growth and development of seedlings. They were also designed to maximize contact between phenolic acids supplied to the systems and seedlings roots. Observed response times for initial and maximum development of inhibition by phenolic acids and the initial and full recovery after phenolic acid depletion were thus potentially maximized although some might argue that optimizing growth environments for seedlings may lead to more robust and less sensitive seedlings.

c. Theoretically any organic compound at the right concentration can be stimulatory, inhibitory or neutral for a given process being measured (e.g., growth, physiology). Soils contain complex mixtures of available/active organic compounds other than phenolic acids. The role of such organic compounds other than phenolic acids have been minimized or in some cases for all intents and purposes eliminated (e.g., in nutrient culture) in our model systems. The enrichment of systems with glucose, methionine, etc. in addition to phenolic acids in several bioassays was an attempt to address this issue.
d. A- and B-horizon Cecil, Portsmouth, or White Store soils used in the model systems represented a range of soils and environments in the Piedmont and Coastal Plain of North Carolina. Here as well, the processes (e.g., reversible sorption and desorption, and irreversible sorption) observed were consistent from soil to soil and varied only in magnitude. However, the biological and physicochemical nature of the soils collected from the field were modified by air drying, sieving, storage, by mixing soil materials with sand for seedling bioassays, and by the addition of nutrients, water, and/or phenolic acids and other carbon sources.

   Air drying, sieving, and storage of soils would, for example, modify soil hydration, soil structure, and microbial populations. The addition of sand (2 units of sand to 1 unit of soil by weight) for the seedling bioassays influenced soil compaction, soil aeration, and reduced the surface areas for sorption of phenolic acids per unit of substrate. These changes would largely modify the magnitude of processes (e.g., seedling growth, phenolic acid uptake by roots, microbial utilization of phenolic acids, and sorption of phenolic acids) and not the actual processes themselves. The primary benefit of such soil manipulations was a reduction in heterogeneity of soils for experimental units within and among experiments. In practical terms the benefits were a consistency within and among experiments and the requirement of fewer replicates to establish patterns or trends.
e. For some sorption and microbial utilization studies soils were sterilized by autoclaving. Sterilizing soils by autoclaving will modify soil physicochemistry and thus possibly alter sorption and microbial utilization of phenolic acids after soils are inoculated (Salonius et al. 1967; Lopes and Wollum 1976; Williams-Linera and Ewel 1984; Dalton et al. 1989a). The most dramatic change observed by Dalton et al. (1989a) for autoclaved Cecil and Portsmouth A and B soils was an increase in Mn. Manganese in autoclaved soils was approximately 2 times higher in A-horizon soils and up to 8 times higher for B-horizons soils. Since this is a multivalent cation, reversible sorption of phenolic acids would very likely be higher for autoclaved soils. Recovery of available ferulic acids by Mehlich III shortly after addition to Cecil and Portsmouth soils was very similar between sterile (i.e., autoclaved) and non-sterilized soils. From a biological perspective the following can be stated: (a) Data from Salonius et al. (1967) suggests that microbial growth of some species may be initially delayed and somewhat suppressed in autoclaved soil, and (b) Williams-Linera and Ewel (1984) observed that 4 tropical species and two cultivars (i.e., soybean and radish) grew slightly better in unsterilized soil and two other tropical species grew slightly better in autoclaved soil, and

f. Enrichment of soils with phenolic acids enhances induction/selection of microorganisms that utilized phenolic acids as a carbon source. This enhancement was particularly dramatic in model systems under laboratory conditions where environmental conditions were particularly suitable for microbial metabolism and reproduction. Microbial populations including phenolic acid-utilizing microbes in laboratory soils and the rhizosphere of seedlings were almost certainly different from field soils. However, since species of microorganisms in soil are very diverse (e.g., perhaps as many as 10^4–10^6 bacterial species per gram of soil) and only a small fraction of these microorganisms can actually be cultured (e.g., < 1% for bacteria) determining such differences can be very difficult, if not impossible (Sylvia et al. 2004). Thus when determining colony-forming units per gram of soil (CFU) for laboratory soils, it was assumed that the CFUs obtained were representative or at least consistently related to soil microbial population behavior of field soils. However, since in field soils large number of microorganisms spend most of their lives just surviving in quiescent or dormant stages because of environmental limitations (for example, bulk-soil bacteria may be active 3 years out of 30 years; Lavelle and Spain 2001), we suspect that active phenolic acid-utilizing microbial populations were much higher in our laboratory soil systems than in field soils particularly when multiple enrichments or treatments of phenolic acids and other organic compounds were applied. Although there are no field population estimates of active phenolic acid-utilizing microbes in bulk soils or rhizospheres, we did observe lower numbers of viable (i.e., active and inactive) phenolic acid-utilizing microbes in soils and rhizospheres of wheat field plots (Blum 2004).

Based on the observations for and limitations of our model systems, the following should be relevant for field systems. Only the timing and magnitude will vary.

a. The removal of phenolic acids from soil solutions by soil processes (e.g., sorption, microbial utilization, and root uptake) and redistribution of phenolic acids within soil (e.g., leaching, capillary movement, mass flow) in concert with phenolic acid inputs (e.g., enrichment and release of reversibly sorbed phenolic acids) will determine the concentration of phenolic acids in bulk-soil solutions at a given location and point in time within the soil. In natural systems inputs of phenolic acids would, of course, be due to leaching of living plant tissues, exudation and secretion from roots, sloughing of root cells and cell lysis, fragmentation and leaching of organic litter and residues, and/or microbial utilization of organic matter and synthesis of phenolic acids.
b. Contact of roots with phenolic acids will occur when roots grow into areas containing sources of available/active phenolic acids or phenolic acids will reach roots by way of capillary flow and/or "transpirational pull" of soil solutions when they are located adjacent to phenolic acid sources. Since phenolic acids can inhibit root growth direct contact between roots and sources with very inhibitory phenolic acid concentrations may be limited.

c. The rhizoplane and rhizosphere are important regulators of phenolic acid contact with root cell membranes, i.e., regulators of primary effects of phenolic acids.
d. Available (i.e., "free" and reversible sorbed) and active phenolic acids (i.e., "free" forms that actually interact with root surfaces) in soils or rhizosphere/rhizoplane will not necessarily be identical. Active phenolic acids are a subset of available phenolic acids.
e. Available/active phenolic acid concentrations in bulk soil may or may not be related to available/active phenolic acid concentrations in the rhizosphere/rhizoplane.
f. Phytotoxicity or stimulation (hormesis) of available/active phenolic acids will not only be regulated by sorption, microbial utilization, root uptake, and leaching but also by a host of environmental factors, e.g., aeration, pH, cations, and types of other organic compounds present.
g. Effects of active phenolic acids on seedling processes are local and not systemic and thus readily reversible given sufficient time and adequate growth conditions. Understanding how roots and sources of phenolic acids (e.g., organic residues) are distributed in soils is essential.
h. Soils, however, do not only contain phenolic acids, but also an array of other organic and inorganic compounds. Since any given compound or ion can stimulate, have no effects, or inhibit a given seedling process depending on its physicochemical state and concentration, and the sensitivity of the seedling being impacted, the effects of active phenolic acids on seedlings will not be independent of other active organic and inorganic compounds present. In fact, the inhibition of phenolic acids, particularly in soils, will really be a product of the action of a set of active organic and inorganic compounds, including organic breakdown products (e.g., breakdown products of phenolic acids), that act individually or in combination as promoters, modifiers, or inhibitors depending on their individual concentrations and their environment, and
i. Except for very unusual circumstances, it is unlikely that estimates of available phenolic acids in field soils based on soil extracts will be directly related to seedling inhibition under field conditions. Soil extracts provide estimates of net or residual concentrations in the soil but provide little, if any, data on the phenolic acids that move through the system in between extraction times.

Finally, effects observed in soil model systems tend to more closely approximate what happens in nature (within limits of course, see above; actually represent soil modified responses), while effects observed in nutrient culture tend to represent maximum potential effects of phenolic acids for a given environment and a given species. Both bioassay systems (nutrient culture and soil culture), therefore, supply useful information about the potential role and the mechanisms of action of phenolic acids but cannot provide information about what is actually happening under field conditions (Inderjit and Dakshini 1995; Romeo and Weidenhamer 1998; Inderjit and Weston 2000; Inderjit and Callaway 2003; Inderjit and Nilsen 2003; Blum 2007).

Why not? Simply put, no matter how hard we try, laboratory environments cannot duplicate the dynamic and heterogeneous conditions that occur in nature nor can they supply actual quantitative data for individual processes under field conditions. At best laboratory model systems can provide some insight into the range of variation, response times, changes in magnitude as the environment changes, interactions between processes, and the overall importance of individual processes to plant–plant allelopathic interactions. To determine what actually happens under field conditions requires field observations and/or experimentation. Here we are limited to either identifying relationships or correlations (e.g., reduction in growth of species A in the presence of species B when potential soil inhibitors are present and released from B) or experimentally manipulating the field environment (i.e., larger scale bioassays). Unfortunately relationships or correlations do not necessarily represent cause and effect. For example, it is very likely that mechanisms of negative interactions between species change over time with changing environmental conditions, e.g., at one moment they may be competing for limited resources (i.e., resource competition) and at another moment they may be interacting by chemical messages (i.e., allelopathy). Add to that the time delays for secondary effects, such as reductions of seedling growth, and colinearities between environmental factors, and the difficulty of identifying cause and effect relationships by simple observations or correlations in nature become all too obvious. Observations and correlations, however, can provide testable hypothesis that can be approached experimentally both in the laboratory and the field.

2.6.1 Promoters, Modifiers, and Inhibitors

On the surface the following statement (see h. Section 2.6) seems straight forward and not particularly noteworthy: "The inhibition of phenolic acids, particularly in soils, will really be a product of the action of a set of active organic and inorganic compounds, including organic breakdown products (e.g., breakdown products of phenolic acids), that act individually or in combination as promoters, modifiers, or inhibitors depending on their individual concentrations and their environment." However, after some reflection one quickly reaches the conclusion that the statement is actually quite noteworthy, although convoluted, and that the underlying assumptions and implications of this statement to field studies are enormous. What then are some of the underlying assumptions and implications?

a. Theoretically any organic or inorganic compound can be stimulatory, neutral, or inhibitory depending on its concentration and physicochemical state and the sensitivity of a seed and seedling based on their genetic makeup and their past and present environments.
b. That the stimulation or inhibition of seed germination or seedling emergence and growth is due to the action of a complex set of interacting environmental variables not just promoters or inhibitors.

c. That modifiers in the soil can turn promoters to neutral compounds and/or inhibitors and inhibitors to neutral compounds and/or promoters by regulating their concentrations and/or physicochemical states.
d. That modifiers of promoters or inhibitors in the soil can be:

  1. neutral organic compounds, and/or other promoters and inhibitors,
  2. inorganic compounds, such as nitrate and multivalent cations,
  3. soil physical characteristics, such as pH, temperature, aeration, and soil moisture,
  4. soil processes, such as sorption, and leaching,
  5. biotic processes such as microbial utilization and synthesis and root exudation, secretions, and uptake,
  6. weather, such as rainfall events, and
  7. various management practices.

e. That many of the modifiers are related and/or interrelated with each other, i.e., a change in one causes other modifiers to be change as well, and
f. That many of these modifiers actually also impact seed germination and/or seedling growth directly.

For field systems, such as wheat no-till (see Chapter 3), where the primary source of organic and inorganic compounds, including allelopathic agents, come from living plants, plant residues, and/or soil organic matter most, if not all, of these assertions are likely to be true. Data from our model system studies clearly lend support to many of these assertions. Thus identifying potential allelopathic agents in plant tissues/residues, in soil organic matter, and/or in soils associated with putative allelopathic plants that stimulate or inhibit seedlings is only one step in determining whether plant–plant allelopathic interactions are operational within a system. Another step is placing promoters and/or inhibitors in the proper context of promoter/modifier/inhibitor complexes within the soil and determining the fluxes of promoters, modifiers, and inhibitors over time. Residual or net concentrations at any point in time are inadequate.

References

Balke NE (1985) Effects of allelochemicals on mineral uptake and associated physiological processes. In: Thompson AC (ed) The chemistry of allelopathy. Biochemical interactions among plants. ACS symposium series, vol 268. American Chemical Society, Washington, DC, pp 161–178

Barkosky RR, Einhellig FA (1993) Effects of salicylic acid on plant-water relationships. J Chem Ecol 19:237–247

Bates-Smith EC (1956) The commoner phenolic constituents of plants and their systematic distribution. Proc R Dublin Sci 27:165–176

Baziramakenga R, Leroux GD, Simard RR (1995) Effects of benzoic and cinnamic acids on membrane permeability of soybean roots. J Chem Ecol 21:1271–1285

Belz RG (2008) Stimulation versus inhibition – bioactivity of parthenin, a phytochemical from *Parthenium hysterophorus* L. Int Dose-Response Soc 6:80–96

Belz RG, Velini ED, Duke SO (2007) Dose/response relationships in allelopathy research. In: Fujii Y, Hiradate S (eds) Allelopathy: new concepts and methodologies. Science Publishers, Enfield, NH, pp 3–29

Bergmark CL, Jackson WA, Volk RJ, Blum U (1992) Differential inhibition by ferulic acid of nitrate and ammonium uptake in *Zea mays* L. Plant Physiol 98:639–645

Bertin C, Yang X, Weston LA (2003) The role of root exudates and allelochemicals in the rhizosphere. Plant Soil 256:67–83

Black RLB, Dix NJ (1976) Utilization of ferulic acid by microfungi from litter and soil. Trans Br Mycol Soc 66:313–317

Blum U (1996) Allelopathic interactions involving phenolic acids. J Nematol 28:259–267

Blum U (1997) The benefits of citrate over EDTA for extracting phenolic acids from soils and plant debris. J Chem Ecol 23:347–362

Blum U (1998) Effects of microbial utilization of phenolic acids and their phenolic acid breakdown products on allelopathic interactions. J Chem Ecol 24:685–708

Blum U (2004) Fate of phenolic allelochemicals in soils: the role of soil and rhizosphere microorganisms. In: Macías FA, Galindo JCG, Molinillo JMG, Cutler HG (eds) Allelopathy: chemistry & modes of action of allelochemicals. CRC Press, Boca Raton, FL, pp 57–76

Blum U (2006) Allelopathy: a soil system perspective. In: Reigosa MJ, Pedrol N, González L (eds) Allelopathy. A physiological process with ecological implications. Springer, Dordrecht, pp 299–340

Blum U (2007) Can data derived from field and laboratory bioassays establish the existence of allelopathic interactions in nature? In: Fujii Y, Hiradate S (eds) Allelopathy: new concepts and methodology. Science Publishers, Enfield, NH, pp 31–38

Blum U, Austin MF, Shafer SR (1999a) The fates and effects of phenolic acids in a plant-microbe-soil model system. In: Macías FA, Galindo JCG, Molinillo JMG, Cutler HG (eds) Recent advances in allelopathy I. A science for the future. Cádiz University Press, Puerto Real, pp 159–166

Blum U, Dalton BR (1985) Effects of ferulic acid, an allelopathic compound, on leaf expansion of cucumber seedlings grown in nutrient culture. J Chem Ecol 11:279–301

Blum U, Dalton BR, Rawlings JO (1984) Effects of ferulic acid and some of its microbial metabolic products on the radicle growth of cucumber. J Chem Ecol 10:1169–1191

Blum U, Dalton BR, Shann JR (1985a) Effects of various mixtures of ferulic acid and some of its microbial metabolic products on cucumber leaf expansion and dry matter in nutrient culture. J Chem Ecol 11:619–641

Blum U, Dalton BR, Shann JR (1985b) Effects of ferulic and *p*-coumaric acids in nutrient culture on cucumber leaf expansion as influenced by pH. J Chem Ecol 11:1567–1582

Blum U, Gerig TM (2005) Relationships between phenolic acid concentrations, transpiration, water utilization, leaf area expansion, and uptake of phenolic acids: nutrient culture studies. J Chem Ecol 31:1907–1932

Blum U, Gerig TM (2006) Interrelationships between *p*-coumaric acid, evapotranspiration, soil water content, and leaf expansion. J Chem Ecol 32:1817–1834

Blum U, Gerig TM, Weed SB (1989) Effects of mixtures of phenolic acids on leaf area expansion of cucumber seedlings grown in different pH Portsmouth A_1 soil materials. J Chem Ecol 15: 2413–2423

Blum U, Gerig TM, Worsham AD, Holappa LD, King LD (1992) Allelopathic activity in wheat-conventional and wheat-no-till soils: development of soil extract bioassays. J Chem Ecol 18:2191–2221

Blum U, Gerig TM, Worsham AD, King LD (1993) Modification of allelopathic effects of *p*-coumaric acid on morning-glory seedling biomass by glucose, methionine, and nitrate. J Chem Ecol 19:2791–2811

Blum U, Heck WW (1980) Effects of acute ozone exposures on snap bean at various stages of its life cycle. Environ Exp Bot 20:73–85

Blum U, King LD, Gerig TM, Lehman ME, Worsham AD (1997) Effects of clover and small grain cover crops and tillage techniques on seedling emergence of some dicotyledonous weed species. Am J Altern Agric 12:146–161

Blum U, Rebbeck J (1989) The inhibition and recovery of cucumber roots given multiple treatments of ferulic acid in nutrient culture. J Chem Ecol 15:917–928

Blum U, Shafer SR (1988) Microbial populations and phenolic acids in soils. Soil Biol Biochem 20:793–800

Blum U, Shafer SR, Lehman ME (1999b) Evidence for inhibitory allelopathic interactions involving phenolic acids in field soils: concepts vs. an experimental model. Crit Rev Plant Sci 18:673–693

Blum U, Staman KL, Flint LJ, Shafer SR (2000) Induction and/or selection of phenolic acids-utilizing bulk-soil and rhizosphere bacteria and their influence on phenolic acid phytotoxicity. J Chem Ecol 26:2059–2078

Blum U, Weed SB, Dalton BR (1987) Influence of various soil factors on the effects of ferulic acid on leaf expansion of cucumber seedlings. Plant Soil 98:111–130

Blum U, Wentworth TR, Klein K, Worsham AD, King LD, Gerig TM, Lyu S-W (1991) Phenolic acid content of soils from wheat-no till, wheat-conventional till, and fallow-conventional till soybean cropping systems. J Chem Ecol 17:1045–1068

Blum U, Worsham AD, King LD, Gerig TM (1994) Use of water and EDTA extractions to estimate available (free and reversibly bound) phenolic acids in Cecil soils. J Chem Ecol 20:341–359

Bonner J (1946) Further investigation of toxic substances which arise from guayule plants: relation of toxic substances to growth of guayule in soil. Bot Gaz 107:343–351

Bonner J, Galston AW (1944) Toxic substances from culture media of guayule which may inhibit growth. Bot Gaz 106:185–198

Booker FL, Blum U, Fiscus EL (1992) Short-term effects of ferulic acid on ion uptake and water relations in cucumber seedlings. J Exp Bot 43:649–655

Caballeira A, Cuervo A (1980) Seasonal variation in allelopathic potential of soil from *Erica australis* L heathland. Acta Oecol-Oecol Plant 1:345–353

Cataldo DA, Schrader LE, Young VL (1974) Analysis by digestion and colorimetric assay of total nitrogen in plant tissue high in nitrates. Crop Sci 14:854–856

Cecchi AM, Koskinen WC, Cheng HH (2004) Sorption-desorption of phenolic acids as affected by soil properties. Biol Fertil Soils 39:235–242

Cheng HH, Haider K, Harper SS (1983) Catechol and chlorocatechols in soil: degradation and extractability. Soil Biol Biochem 15:311–317

Chiou CT (1989) Theoretical consideration of the partition uptake of nonionic organic compounds by soil organic matter. In: Sawhney BK (ed) Reactions and movement of organic chemicals in soils, SSSA Special Publication 22. Soil Science Society of America, Madison, WI, pp 1–29

Chou CH, Patrick ZA (1976) Identification and phytotoxic activity of compounds produced during decomposition of corn and rye residues in soil. J Chem Ecol 2:369–387

Connors KA, Lipari JM (1976) Effects of cycloamyloses on apparent dissociation constants of carboxylic acids and phenols: equilibrium analytical selectivity induced by complex formation. J Pharmaceut Sci 65:379–383

Dagley S (1971) Catabolism of aromatic compounds by microorganisms. Adv Microb Physiol 6:1–42

Dalton BR (1989) Physicochemical and biological processes affect the recovery of exogenously applied ferulic acid from tropical soils. Plant Soil 115:13–22

Dalton BR (1999) The occurrence and behavior of plant phenolic acids in soil environments and their potential involvement in allelochemical interference interactions: methodological limitations in establishing conclusive proof of allelopathy. In: Inderjit, Daskshini KMM, Foy CL (eds) Principles and practices in plant ecology: allelochemical interactions. CRC Press, Boca Raton, FL, pp 57–74

Dalton BR, Blum U, Weed SB (1983) Allelopathic substances in ecosystems: effectiveness of sterile soil components in altering recovery of ferulic acid. J Chem Ecol 9:1185–1201

Dalton BR, Blum U, Weed SB (1989a) Plant phenolic acids in soils: sorption of ferulic acid by soil and soil components sterilized by different techniques. Soil Biol Biochem 21:1011–1018
Dalton BR, Blum U, Weed SB (1989b) Differential sorption of exogenously applied ferulic, *p*-coumaric, *p*-hydroxybenzoic, and vanillic acids in soil. Soil Sci Soc Am J 53:757–762
Dalton BR, Weed SB, Blum U (1987) Plant phenolic acids in soils: a comparison of extraction procedures. Soil Sci Soc Am J 51:1515–1521
Dao TH (1987) Sorption and mineralization of plant phenolic acids in soil. In: Waller GR (ed) Allelochemicals: role in agriculture and forestry. ACS symposium series, vol 330. American Chemical Society, Washington, DC, pp 358–370
Doetsch RN, Cook TM (1973) Introduction of Bacteria and their ecobiology. University Park Press, Baltimore, MD
Downs RJ, Thomas JF (1983) Phytotron procedural manual for controlled environment laboratories. North Carolina Agric Exper Station Tech Bull No. 244 (revised)
Duke SO, Cedergreen N, Velini ED, Belz RG (2006) Hormesis: is it an important factor in herbicide use and allelopathy? Outlook on Pest Manag February 2006:29–33
Duke SO, Williams RD, Markhart AH (1983) Interaction of moisture stress and three phenolic compounds on lettuce seed germination. Ann Bot 52:923–926
Einhellig FA (1986) Mechanisms and modes of action. In: Putnam AR, Tang CS (eds) The science of allelopathy. Wiley, New York, NY, pp 171–188
Einhellig FA (1987) Interactions among allelochemicals and other stress factors of the plant environment. In: Waller GR (ed) Allelochemicals: role in agriculture and forestry. ACS symposium series, vol 330. American Chemical Society, Washington, DC, pp 343–357
Einhellig FA (1996) Interactions involving allelopathy in cropping systems. Agron J 88:886–893
Einhellig FA (1999) An integrated view of allelochemicals amid multiple stresses. In: Inderjit, Dakshini KMM, Foy CL (eds) Principles and practices in plant ecology: allelochemical interactions. CRC Press, Boca Raton, FL, pp 479–494
Einhellig FA (2002) The physiology of allelochemical action: clues and views. In: Reigosa MJ, Pedrol N (eds) Allelopathy: from molecules to ecosystems. Science Publishers Inc, Einfield, NH, pp 1–23
Einhellig FA, Eckrich PC (1984) Interaction of temperature and ferulic acid stress on grain sorghum and soybean. J Chem Ecol 10:161–170
Einhellig FA, Kuan L (1971) Effects of scopoletin and chlorogenic acid on stomatal aperture in tobacco and sunflower. Bull Torrey Bot Club 98:155–162
Einhellig FA, Muth MS, Schon MK (1985) Effects of allelochemicals on plant-water relationships. In: Thompson AC (ed) The chemistry of allelopathy: biochemical interactions among plants. ACS symposium series, vol 268. American Chemical Society, Washington, DC, pp 179–195
Einhellig FA, Rice EL, Risser PG, Wender SH (1970) Effects of scopoletin on growth, CO_2 exchange rates, and concentrations of scopoletin, scopolin, and chlorogenic acid in tobacco, sunflower and pigweed. Bull Torrey Bot Club 97:22–33
Evans WC (1963) The microbial degradation of aromatic compounds. J Gen Microbiol 32: 177–185
Fisher K (2002) Removal of heavy metals from soil components and soil by natural chelating agents. Part I: displacement from clay minerals and peat by L-cysteine and L-penecillamine. Water Air Soil Pollut 137:267–286
Fisher K, Bipp HP (2002) Removal of heavy metals from soil components and soil by natural chelating agents. Part II: soil extraction by sugar acids. Water Air Soil Pollut 138:271–288
Flaig W (1971) Organic compounds in soil. Soil Sci 111:19–33
Fry CF (1988) The growing plant cell wall: chemical and metabolic analysis. Longman Scientific and Technical, Harlow
Gerig TM, Blum U (1991) Effects of mixtures of four phenolic acids on leaf area expansion of cucumber seedlings grown in Portsmouth B1 soil materials. J Chem Ecol 17:29–40
Gerig TM, Blum U (1993) Modification of an inhibition curve to account for effects of a second compound. J Chem Ecol 19:2783–2790

Gerig TM, Blum U, Meier K (1989) Statistical analysis of the joint inhibitory action of similar compounds. J Chem Ecol 15:2403–2412
Glass ADM (1973) Influence of phenolic acids on ion uptake. I. Inhibition of phosphate uptake. Plant Physiol 51:1037–1041
Glass ADM (1974) Influence of phenolic acids on ion uptake. III. Inhibition of potassium uptake. J Exp Bot 25:1104–1113
Glass ADM (1975) Inhibition of phosphate uptake in barley roots by hydroxy-benzoic acids. Phytochemistry 14:2117–2130
Glass ADM (1976) The allelopathic potential of phenolic acids associated with the rhizosphere of *Pteridium aquilinum*. Can J Bot 54:2440–2444
Glass ADM, Dunlop J (1974) Influence of phenolic acids on ion uptake. IV. Depolarization of membrane potentials. Plant Physiol 54:855–858
Goodwin TW, Mercer EI (1983) Introduction to plant biochemistry, 2nd edn. Pergaman Press, Oxford
Greenland DJ (1965) Interactions between clays and organic compounds. Part I. Mechanisms of interactions between clays and defined organic matter. Soil Fertil 28:415–425
Greenland DJ (1971) Interactions between humic and fulvic acids and clays. Soil Sci 111:34–41
Guenzi WD, McCalla TM (1966) Phytotoxic substances extracted from soil. Soil Sci Soc Am Proc 30:214–216
Haider K, Martin JP (1975) Decomposition of specifically carbon-14 labeled benzoic and cinnamic acid derivatives in soil. Soil Sci Soc Am Proc 39:657–662
Haider K, Martin JP, Rietz E (1977) Decomposition in soil of ^{14}C-labeled coumaryl alcohols; free and linked into dehydropolymer and plant lignins and model humic acids. Soil Sci Soc Am J 41:556–562
Hall AB, Blum U, Fites RC (1982) Stress modification of allelopathy of *Helianthus annuus* L. debris on seed germination. Am J Bot 69:776–783
Hall AB, Blum U, Fites RC (1983) Stress modification of allelopathy of *Helianthus annuus* L. debris on seedling biomass production of *Amaranthus retroflexus* L. J Chem Ecol 9:1213–1222
Harborne JB (1982) Introduction to ecological biochemistry. Academic Press, London
Harborne JB (1984) Phytochemical methods. A guide to modern plant analysis. Chapman and Hall, New York, NY
Harborne JB (1990) Plant phenolics. In: Bell EA, Charlwood BV (eds) Secondary plant products. Springer, Berlin, pp 331–401
Harder W, Dijkhuizen L (1982) Strategies of mixed substrate utilization in microorganisms. Philos Trans R Soc Lond 297:459–480
Harper JR, Balke NE (1981) Characterization of the inhibition of K^+ absorption in oat roots by salicylic acid. Plant Physiol 68:1349–1553
Harris RK, Sommers LE (1968) Plate-dilution frequency technique for assay of microbial ecology. Appl Microbiol 16:330–334
Hartel PG (1998) The soil habitat. In: Sylvia DM, Fuhrmann JJ, Hartel PG, Zuberer DA (eds) Principles and applications of soil microbiology. Prentice Hall Inc., Upper Saddle River, NJ, pp 21–43
Hasset JJ, Banwart WL (1989) The sorption of nonpolar organics by soils and sediments. In: Sawhney BL, Brown K (eds) Reactions and movement of organic chemicals in soils, SSSA Special Publication 22. Science Society of America, Madison, WI, pp 31–44
Henderson MEK (1956) A study of the metabolism of phenolic compounds by soil fungi using spore suspension. J Gen Microbiol 14:684–691
Henderson MEK, Farmer VC (1955) Utilization by soil fungi of *p*-hydroxybenzaldehyde, ferulic acid, syringaldehyde, and vanillin. J Gen Microbiol 12:37–46
Henry G (1985) Allelopathy: the harmful effects of chemicals produced by one plant upon another, 1979–85 (335 citations). US Dept of Agriculture, National Agricultural Libarary, Beltsville
Henry G (1987) Allelopathy: the effects of chemicals produced by plants, 1982–86 (248 citations). US Dept of Agriculture, National Agricultural Libarary, Beltsville

Hoagland DR, Arnon DJ (1950) The water-culture method of growing plants without soil. Calif Agric Exp Sta Circ 347
Holappa LD, Blum U (1991) Effects of exogenously applied ferulic acid, a potential allelopathic compound, on leaf growth, water utilization, and endogenous abscisic acid levels of tomato, cucumber, and bean. J Chem Ecol 17:865–886
Huang PM, Wang MC, Wang MK (1999) Catalytic transformation of phenolic compounds in the soil. In: Inderjit, Dakshini KMM, Foy CL (eds) Principles and practices in plant ecology: allelochemical interactions. CRC Press, Boca Raton, FL, pp 287–306
Huang PM, Wang TSC, Wang MK, Wu MH, Hsu NW (1977) Retention of phenolic acids by noncrystalline hydroxy-aluminum and -iron compounds and clay minerals in soil. Soil Sci 123:213–219
Hughes GR, Averre CW, Sorensen KA (1983) Growing pickling cucumbers in North Carolina. AG 315. NC Agric Ext Serv Bull, Raleigh. North Carolina Cooperative Extension Service, Raleigh NC
Inderjit (2005) Soil microorganisms: an important determinant of allelopathic activity. Plant Soil 274:227–236
Inderjit, Bhowmik PC (2004) Sorption of benzoic acid onto soil colloids and its implications for allelopathy studies. Biol Fertil Soils 40:345–348
Inderjit, Callaway RM (2003) Experimental designs for the study of allelopathy. Plant Soil 256: 1–11
Inderjit, Dakshini KMM (1995) On laboratory bioassays in allelopathy. Bot Rev 61:28–44
Inderjit, Nilsen ET (2003) Bioassays and field studies for allelopathy in terrestrial plants: progress and problems. Crit Rev Plant Sci 22:221–238
Inderjit, Streibig JC, Olofsdotter M (2002) Joint action of phenolic acids mixtures and its significance in allelopathy research. Physiol Plant 114:422–428
Inderjit, Weston LA (2000) Are laboratory bioassays for allelopathy suitable for prediction of field responses? J Chem Ecol 26:2111–2118
Kaminsky R, Muller WH (1977) The extraction of soil phytotoxins using neutral EDTA solution. Soil Sci 124:205–210
Kaminsky R, Muller WH (1978) A recommendation against the use of alkaline soil extractions in the study of allelopathy. Plant Soil 49:641–645
Katase T (1981a) Stereoisomerization of *p*-coumaric and ferulic acid during incubation in peat soil extract solution by exposure to fluorescent light. Soil Sci Plant Nutr 27:421–427
Katase T (1981b) The different forms in which *p*-hydroxybenzoic, vanillic, and ferulic acids exist in a peat soil. Soil Sci 132:436–443
Kenttamaa J, Raisanen S, Auterinen L, Lindberg JJ (1970) Ionization constants of coniferyl alcohol and related phenols. Suom Kemistilehti B 43:333–336
Klein K, Blum U (1990a) Inhibition of cucumber leaf expansion by ferulic acid in split-root experiments. J Chem Ecol 16:455–463
Klein K, Blum U (1990b) Effects of soil nitrogen level on ferulic acid inhibition of cucumber leaf expansion. J Chem Ecol 16:1371–1383
Kunc F (1971) Decomposition of vanillin by soil microorganisms. Folia Microbiol 16:41–50
Lavelle P, Spain AV (2001) Soil ecology. Kluwer Academic Publisher, Dordrecht
Leben C (1961) Microorganisms on cucumber seedlings. Phytopathology 51:533–557
Lehman ME, Blum U (1997) Cover crop debris effects on weed emergence as modified by environmental factors. Allelopathy J 4:69–88
Lehman ME, Blum U (1999a) Influence of pretreatment stresses on inhibitory effects of ferulic acid, an allelopathic phenolic acid. J Chem Ecol 25:1517–1529
Lehman ME, Blum U (1999b) Evaluation of ferulic acid uptake as a measurement of allelochemical dose: effective concentration. J Chem Ecol 25:2585–2600
Lehman ME, Blum U, Gerig TM (1994) Simultaneous effects of ferulic and *p*-coumaric acids on cucumber leaf expansion in split-root experiments. J Chem Ecol 20:1773–1782

Lehmann RG, Cheng HH, Harsh JB (1987) Oxidation of phenolic acids by soil iron and manganese oxides. Soil Sci Soc Am J 51:352–356

Leo AJ, Hansch C, Elkins D (1971) Partition coefficients and their uses. Chem Rev 71:525–616

Lindsay WL, Norwell WA (1978) Development of a DTPA soil test for zinc, manganese, and copper. Soil Sci Soc Am J 42:421–428

Littell RC, Milliken GA, Stroup WW, Wolfinger RD (1996) SAS system for mixed models. SAS Institute, Cary, NC

Lodhi MAK (1978) Allelopathic effects of decaying litter on dominant trees and their associated soil in a lowland forest community. Am J Bot 65:340–344

Lopes AS, Wollum AG (1976) Comparative effects of methyl-bromide, propylene oxide, and autoclave sterilization of specific soil chemical characteristics. Turrialba 26:351–355

Lowe RH, Hamilton JL (1967) Rapid method for determination of nitrate in plant and soil extracts. Agric Food Chem 15:359–361

Lyu S-W, Blum U (1990) Effects of ferulic acid, an allelopathic compound, on net P, K, and water uptake by cucumber seedlings in a split-root system. J Chem Ecol 16:2429–2439

Lyu S-W, Blum U, Gerig TM, O'Brien TE (1990) Effects of mixtures of phenolic acids on phosphorus uptake by cucumber seedlings. J Chem Ecol 16:2559–2567

Makino T, Takahashi Y, Sakurai Y, Nanzyo M (1996) Influence of soil chemical properties on adsorption and oxidation of phenolic acids in soil suspension. Soil Sci Plant Nutr 42: 867–879

Marschner H, Römheld V (1996) Root-induced changes in the availability of micronutrients in the rhizosphere. In: Waisel Y, Eshel A, Kafkafi U (eds) Plant roots: the hidden half, 2nd edn. Marcel Dekker Inc, New York, NY

Martin JP, Haider K (1971) Microbial activity in relation to soil humus formation. Soil Sci 111: 54–63

Martin JP, Haider K (1976) Decomposition of specifically carbon-14-labeled ferulic acid: free and linked into model humic acid-type polymers. Soil Sci Soc Am J 40:377–380

Martin JP, Haider K (1979) Effects of concentration on decomposition of some ^{14}C-labeled phenolic compounds, benzoic acid, glucose, wheat straw, and *Chlorella* protein in soil. Soil Sci Soc Am J 43:917–920

Martin JP, Haider K, Wolf D (1972) Synthesis of phenolic polymers by *Hedersonula toruloidea* in relation to humic acid formation. Soil Sci Soc Am Proc 36:311–315

Moreland DE, Novitzky WP (1987) Effects of phenolic acids, coumarins, and flavonoids on isolated chloroplasts and mitochondria. In: Waller GR (ed) Allelochemicals: role in agriculture and forestry. ACS symposium series, vol 330. American Chemical Society, Washington, DC, pp 247–261

Mundt JO, Hinkle NF (1976) Bacteria within ovules and seeds. Appl Environ Microbiol 32: 694–697

Nagarajah S, Posner AM, Quirk JP (1970) Competitive adsorption of phosphate with polygalactoronic and other organic anions on kaolinite and oxide surfaces. Nature (Lond) 228:83–85

Nordstrom CG, Lindberg JJ (1965) Thermodynamic ionization constants of phenolic corboxylic acid related to cinnamic acid and lignin. Suom Kemistilehti 38:291–295

Ohno T (2001) Oxidation of phenolic acid derivatives by soil and its relevance to allelopathic activity. J Environ Qual 30:1631–1635

Ohno T, First PR (1998) Assessment of the Folin and Ciocalteu's method for determining soil phenolic carbon. J Environ Qual 27:776–782

Ohno T, Horesh MY, Merrit KA, Wagai R (2002) Calcium and pH effects on salicylic acid phytotoxicity. Allelopathy J 9:19–25

Pandey DK (1994) Inhibition of Salvinia (*Salvinia molesta* Mitchell) by parthenium (*Parthenium hysterophorus* L.). II. Relative effect of flower, leaf, stem, and root residue on salvinia and paddy. J Chem Ecol 20:3123–3131

Papanastasiou AC (1982) Kinetics of biodegradation of 2,4-dichlorophenoxyacetate in the presence of glucose. Biotechnol Bioeng 24:2001–2011

Parfitt RL, Farmer VC, Russell JD (1977) Adsorption on hydrous oxides I. Oxalate and benzoate on goethite. J Soil Sci 28:29–39
Patrick ZA (1971) Phytotoxic substances associated with the decomposition in the soil of plant residue. Soil Sci 111:13–18
Patterson DT (1981) Effects of allelopathic chemicals on growth and physiological responses to soybean (*Glycine max*). Weed Sci 29:53–58
Paul EA, Clark FE (1989) Soil microbiology and biochemistry. Academic Press, San Diego, CA
Politycka B (1998) Phenolics and the activities of phenylalanine ammonia-lyase, phenol-ß-glucosyltransferase and ß-glucosidase in cucumber roots as affected by phenolic allelochemicals. Acta Physiol Plant 20:405–410
Politycka B, Kozlowska M, Mielcarz B (2004) Cell wall peroxidases in cucumber roots induced by phenolic allelochemicals. Allelopathy J 13:29–35
Politycka B, Mielcarz B (2007) Involvement of ethylene in growth of cucumber roots by ferulic and *p*-coumaric acids. Allelopathy J 19:451–460
Pue KJ, Blum U, Gerig TM, Shafer SR (1995) Mechanism by which noninhibitory concentrations of glucose increase inhibitory activity of *p*-coumaric acid on morning-glory seedling biomass accumulation. J Chem Ecol 21:833–847
Putnam AR, Tang CS (1986) The science of allelopathy. Wiley, New York, NY
Qu XH, Wang JG (2008) Effect of amendments with different phenolic acids on soil microbial biomass, activity, and community diversity. Appl Soil Ecol 39:172–179
Radford PJ (1967) Growth analysis formulae-their use and abuse. Crop Sci 7:171–175
Rice EL (1974) Allelopathy. Academic Press, Orlando, FL
Rice EL (1979) Allelopathy-an update. Bot Rev 45:15–109
Rice EL (1983) Pest control with nature's chemicals: allelochemics and pheromones in gardening and agriculture. University of Oklahoma Press, Norman, OK
Rice EL (1984) Allelopathy. Academic Press, Orlando, FL
Rice EL (1986) Allelopathic growth stimulation. In: Putnam AR, Tang CS (eds) The science of allelopathy. Wiley, New York, NY, pp 23–42
Robinson T (1967) The organic constituents of higher plants. Their chemistry and interrelationships, 2nd edn. Burgess Publishing Company, Minneapolis, MN
Romeo JT, Weidenhamer JD (1998) Bioassays for allelopathy in terrestrial plants. In: Haynes KF, Millar JG (eds) Methods in chemical ecology: bioassay methods. Kluwer Academic Publishing, Norwell, MA, pp 179–209
Rovira AD (1991) Rhizosphere research-85 years of progress and frustration. In: Keister DL, Cregan PB (eds) Beltsville symposium in agricultural research. 14. The rhizosphere and plant growth. Kluwer Academic Press, Dordrecht, pp 3–13
Salonius PO, Robin JB, Chase FE (1967) A comparison of autoclaved and gamma-irradiated soils as media for microbial colonization experiments. Plant Soil 27:239–248
SAS Institute Inc. (1999) SAS/STAT user's guide, version 8. SAS Publishing, Cary, NC
Schmidt SK (1988) Degradation of juglone by soil bacteria. J Chem Ecol 14:1561–1571
Schmidt EL (1991) Methods for microbial autecology in the soil rhizosphere. In: Keister DL, Cregan PB (eds) Beltsville symposium in agricultural research. 14. The rhizosphere and plant growth. Kluwer Academic Press, Dordrecht, pp 81–89
Schmidt SK, Ley RE (1999) Microbial competition and soil structure limit the expression of allelopathy. In: Inderjit, Dakshini KMM, Foy CL (eds) Principles and practices in plant ecology: allelochemical interactions. CRC Press, Boca Raton, FL, pp 339–351
Shafer SR, Blum U (1991) Influence of phenolic acids on microbial populations in the rhizosphere of cucumber. J Chem Ecol 17:369–389
Shafer SR, Blum U, Horton SJ, Hesterberg DL (1998) Biomass of tomato seedlings exposed to an allelopathic phenolic acid and enriched atmospheric carbon dioxide. Water Air Soil Pollut 106:123–136
Shann JR, Blum U (1987a) The uptake of ferulic and *p*-hydroxybenzoic acids by *Cucumis sativus*. Phytochemistry 26:2959–2964

Shann JR, Blum U (1987b) The utilization of exogenously supplied ferulic acid in lignin biosynthesis. Phytochemistry 26:2977–2982

Shindo H, Kuwatsuka S (1976) Behavior of phenolic substances in the decay process of plants. IV. Adsorption and movement of phenolic acids in soils. Soil Sci Plant Nutr 22:23–33

Siqueira JO, Nair MG, Hammerschmidt R, Safir GR (1991) Significance of phenolic compounds in plant-soil-microbial systems. Crit Rev Plant Sci 10:63–121

Sparling GP, Ord BG, Vaughan D (1981) Changes in microbial biomass and activity in soils amended with phenolic acids. Soil Biol Biochem 13:455–460

Sparling GP, Vaughan D (1981) Soil phenolic acids and microbes in relation to plant growth. J Sci Food Agric 32:625–626

Staman K, Blum U, Louws F, Robertson D (2001) Can simultaneous inhibition of seedling growth and stimulation of rhizosphere bacterial populations provide evidence for phytotoxin transfer from plant residues in the bulk soil to the rhizosphere of sensitive species? J Chem Ecol 27: 807–829

Stowe LG, Osborn A (1980) The influence of nitrogen and phosphorus levels on the phytotoxicity of phenolic compounds. Can J Bot 58:1149–1153

Strobel BW (2001) Influence of vegetation on low-molecular-weight carboxylic acids in soil solution – a review. Geoderma 99:169–198

Sugi SF, Schimel JP (1993) Decomposition and biomass incorporation of ^{14}C-labeled glucose and phenolics in taiga forest floor: effect of substrate quality, successional state, and season. Soil Biol Biochem 25:1379–1389

Sylvia DM, Fuhrmann JJ, Hartel PG, Zuberer DA (2004) Principles and application of soil microbiology, 2nd edn. Prentice Hall, Upper Saddle River, NJ

Tack BF, Chapman PJ, Dagley S (1972) Metabolism of gallic and syringic acids by *Pseudomonas putida*. J Biol Chem 247:6438–6443

Taussky HH, Shorr E (1953) A microcolorimetric method for the determination of inorganic phosphorus. J Biol Chem 202:675–685

Tharayil N, Bhowmik PC, Xing B (2006) Preferential sorption of phenolic phytotoxins to soil: implications for altering the availability of allelochemicals. J Agric Food Chem 54: 3033–3040

Tharayil N, Bhowmik PC, Xing B (2008) Bioavailability of alleochemicals as affected by companion compounds in soil matrices. J Agric Food Chem 56:3706–3713

Thompson AC (1985) The chemistry of allelopathy. Biochemical interactions among plants, ACS symposium series, vol 268. American Chemical Society, Washington, DC

Turner JA, Rice EL (1975) Microbial decomposition of ferulic acid in soil. J Chem Ecol 1:41–58

Vaughan D, Sparling GP, Ord BG (1983) Amelioration of the phytotoxicity of phenolic acids by some soil microbes. Soil Biol Biochem 15:613–614

Walker TS, Bais HP, Grotewold E, Vivanco JM (2003) Root exudation and rhizosphere biology. Plant Physiol 132:44–51

Waller GR (1987) Allelochemicals: role in agriculture and forestry. ACS symposium series, vol 330. American Chemical Society, Washington, DC

Wang TSC, Huang PM, Chou C-H, Chen J-H (1986) The role of soil minerals in the abiotic polymerization of phenolic compounds and formation of humic substances. In: Huang PM, Schnitzer M (eds) Interactions of soil minerals with natural organics and microbes, SSSA Special Publication 17. Soil Science Society of America, Madison, WI, pp 251–281

Wang TSC, Song WL, Ferng YL (1978) Catalytic polymerization of phenolic compounds by clay minerals. Soil Sci 126:15–21

Waters ER, Blum U (1987) The effects of single and multiple exposures of ferulic acid on the vegetative and reproductive growth of *Phaseolus vulgaris* BBL-290. Am J Bot 74:1635–1645

Watson JR, Posner AM, Quirk JP (1973) Adsorption of herbicide 2,4-D on goethite. J Soil Sci 24:503–511

Weidenhamer JD, Morton TC, Romeo JT (1987) Solution volume and seed number – often overlooked factors in allelopathic bioassays. J Chem Ecol 13:1481–1491

Whitehead DC, Buchan H, Hartley RD (1979) Composition and decomposition of roots of ryegrass and red clover. Soil Biol Biochem 11:619–628
Whitehead DC, Dibb H, Hartley RD (1982) Phenolic compounds in soil as influenced by the growth of different plant species. J Appl Ecol 19:579–588
Whitehead DC, Dibb H, Hartley RD (1983) Bound phenolic compounds in water extracts of soils, plant roots and leaf litter. Soil Biol Biochem 15:133–136
Williams-Linera G, Ewel JJ (1984) Effects of autoclave sterilization of a tropical andept on seed germination and seedling growth. Plant Soil 82:263–268
Williamson GB, Weidenhamer JD (1990) Bacterial degradation of juglone. Evidence against allelopathy? J Chem Ecol 16:1739–1752
Wolf DC, Dao TH, Scott HD, Lavy TL (1989) Influence of sterilization methods on selected soil microbiological, physical, and chemical properties. J Environ Qual 18:39–44
Yu JQ, Matsui Y (1997) Effects of root exudates of cucumber (*Cucumis sativus*) and allelochemicals on ion uptake by cucumber seedlings. J Chem Ecol 23:817–827
Zanardo DIL, Lima RB, Ferrarese MdeLL, Bubna GA, Ferrarese-Filho O (2009) Soybean root growth inhibition and lignification induced by *p*-coumaric acid. Environ Exp Bot 66:25–30
Zhang Y, Gu M, Xia X, Shi K, Zhou Y, Yu Z (2009) Effects of phenylcarboxylic acids on mitosis, endoreduplication and expression of cell cycle-related genes in roots of cucumber (*Cucumis sativus* L.). J Chem Ecol 35:679–688

Chapter 3
Plant–Plant Allelopathic Interaction. Phase II: Field/Laboratory Experiments

Abstract This chapter describes the underlying criteria and assumption in the development and execution of field and associated laboratory bioassays. It provides details and commentary regarding the materials and methods used. More specifically, describes how glyphosate-desiccated wheat and other cover crops (crimson clover, subterranean clover, and rye) in no-till systems can directly and indirectly influence morningglory, pigweed, and prickly sida seedling emergence, with emphasis on the role of phenolic acids in plant residues and soil.

Simple phenolic acids and their polymers are ubiquitous in plants, plant residues, soil solutions, and soil organic matter (Bates-Smith 1956; Goodwin and Mercer 1983; Rice 1984; Harborne 1990, 1998; Blum et al. 1999b; Blum 2006; Vermerris and Nicholson 2006) and have important roles/functions within and for plants (e.g., synthesis of secondary cell walls, fibers, and pigments, protection against UV radiation, disease resistance, and defense against herbivores; see Harborne 1993; Siqueira et al. 1991; Vermerris and Nicholson 2006) and within and for ecosystems (e.g., species diversity, the succession and productivity of natural and managed ecosystems, and plant–plant allelopathic interactions; see Rice 1983, 1984, 1995; Putnam and Tang 1986; Waller 1987; Rizvi and Rizvi 1992; Inderjit and Mallik 2002; Reigosa and Pedrol 2002; Zeng et al. 2008). However, the importance of the roles/functions of phenolic acids within ecosystems, particularly in regard to plant–plant allelopathic interactions, has been and continues to be questioned (Schmidt 1988; Williamson and Weidenhamer 1990; Schmidt and Ley 1999; Blum 2004).

That simple phenolic acids such as benzoic and cinnamic acid derivatives are released from living plants and plant residues into the soil environment (see Rice 1984; Siqueira et al. 1991), and that such phenolic acids at the right concentrations can influence in a negative and/or positive manner sensitive seedling physiology, growth and development, plant reproduction, and bulk-soil and rhizosphere/rhizoplane microbial populations, is not in question (see Chapter 2). What is being questioned is whether phenolic acids released from one plant taxon and/or its residues can actually influence the germination, growth, and development of other plant taxa within their immediate vicinity. The arguments against such interactions largely revolve around the distribution, mobility, turnover rates, concentrations, and

U. Blum, *Plant–Plant Allelopathic Interactions*, DOI 10.1007/978-94-007-0683-5_3,

phytotoxicity of phenolic acids in soils and the past inability to demonstrate direct relationships between phenolic acids in soils and plant responses under field conditions. Given the proximity of experimental farms to the campus of North Carolina State University, we chose to address some of these areas of concern/debate by characterizing the potential role of phenolic acids in managed cover crop no-till ecosystems, more specifically their potential role in the inhibition of broadleaf weed seedling emergence.

3.1 Annual Broadleaf Weed Control in No-Till Systems

In an effort to implement conservation measures, farmers have experimented with a variety of production methods, including reduced or zero tillage production. In association with the implementation of these production methods there has been an increase in the use of small grain and legume cover crops. Besides erosion control, increased fertility resulting from nutrient recycling, increased organic matter, reduced evapotranspiration, increased water infiltration into soils, reduced use of energy, reduced nitrogen fertilizer with legume cover crops, improved soil physical properties, reduced use of pesticides, reduced nutrient run off, and moderation of summer and winter soil temperatures (see review by Nagabhushana et al. 2001), the use of cover crops and their residues resulted, in a number of instances in early season weed control (Barnes and Putnam 1983; Putnam et al. 1983; Putnam and DeFrank 1983; Einhellig and Leather 1988; Enache and Ilnicki 1988; Teasdale 1988; Worsham 1989, 1990; Nagabhushana et al. 2001; Belz 2007). At present the most promising cover crop residues for annual broadleaf weed control in temperate regions of the world are the small grains such as wheat and rye (Barnes and Putnam 1983; Barnes et al. 1986; Liebl and Worsham 1983; Putnam et al. 1983; Shilling et al. 1985; Worsham 1989, 1990; Yongoing 2005). For example, in North Carolina Worsham (1989) noted that straw management and tilling in no-till planted crops (corn, soybean, sorghum, and tobacco) affected the level of early-season weeds. He found that when compared to no-cover crop tilled plots that:

a. removing wheat or rye straw plus tilling resulted in a 30 and 9% suppression, respectively,
b. removing wheat or rye straw without tilling resulted in 50 and 43% suppression, respectively,
c. removing wheat or rye straw, tilling, and then replacing the wheat or rye straw resulted in a 60% suppression for both, and
d. leaving the wheat or rye straw without tilling resulted in 81 and 76% suppression, respectively, of broadleaf weeds such as red-root pigweed, common lamb's-quarter, common ragweed, morningglory, prickly sida, and sicklepod.

Worsham and the observations of others suggest that a complex set of mechanisms may be involved in the regulation of annual broadleaf weed germination, emergence, and growth by cover crop residues and tillage practices. These may include the following: the physical barrier and shading associated with residues (Fester and Peterson 1979; Worsham 1989), the modification of soil temperature, pH, and soil moisture by residues (Fester and Peterson 1979; Lehman and Blum 1997; Xu et al. 2006), the reduction in soil disturbance and changes in compaction of no-till systems (Worsham 1989), the immobilization of nutrients such as nitrogen (Kimber 1973; Hadas et al. 2004), increased seed predation (Brust and House 1988; Reader 1991; Brust 1994), and the release of inhibitory compounds such as phenolic acids (Guenzi and McCalla 1966; Liebl and Worsham 1983; Shilling et al. 1985; Blum et al. 1991; Wu et al. 2000a, b, 2001a, b; Tapin et al. 2006), organic acids (Patrick 1971; Lynch 1977; Tang and Waiss 1978; Shilling et al. 1985), hydroxamic acids (Willard and Penner 1976; Niemeyer et al. 1989; Wu et al. 2001c; Macías et al. 2005; Mathiassen et al. 2006), volatile compounds (Buttery et al. 1985; Bradow 1991) and/or other compounds (Nakano et al. 2006). How important each of these factors may be in actually regulating annual broadleaf weed germination, emergence, and growth under field conditions, however, is still not entirely clear. Our inability to identify specific causes is partly due to the fact that these factors do not act in isolation but their actions and observed effects are a result of a complex set of interactions (Elliott and Cheng 1987). Although the focus in this chapter is primarily on allelopathic effects of phenolic acids, considerable attention has also been given to other factors (e.g., soil moisture, temperature, soil pH, and nutrition) that may interact and modify the effects of phenolic acids.

3.2 Materials, Methods, and Commentary

3.2.1 Soil and Plant Tissue/Residue Analyses

Cecil soil samples (Typic Hapludults, clayey, kaolinitic, thermic) were taken from no-till and tilled plots in a long-term rotation study of various low-input cropping systems located 5 km south of the North Carolina State University Campus, Raleigh, NC with or without the following living, mature, or desiccated cover crops: wheat (*Triticum aestivum* L. "Coker 983" or "Southern States 555"), rye (*Secale cereale* L. "Abruzzi"), crimson clover (*Trifolium incarnatum* L. "Tibbee"), or subterranean clover (*T. subterranean* L. "Mount Barker"). No crops (e.g., corn, soybean) were planted in these plots (Blum et al. 1997, 2002; for justification see Sections 3.2.3.4 and 3.4.3), but broadleaf weeds were broadcast onto these plots (see Section 3.2.3). For another experiment wheat (*Triticum aestivum* L. "Coker 916") was harvested and for these plots only wheat stubble and roots (not tilled or tilled under) were present at the time of soil sampling. Plots without wheat stubble and roots were also included. Soybeans (*Glycine max* L. "Delapine 417") were planted into the ± wheat stubble plots (Blum et al. 1991, 1992). There were thus three treatments: wheat stubble/soybean (no-till), wheat stubble tilled under/soybean (conventional-till), and fallow/soybean (conventional-till).

Zero to 2.5 cm and/or 0–10 cm soil cores (5.5 cm in diameter) were taken from field plots, sieved (3 mm sieve), placed in plastic bags, frozen, and stored at –20°C (Blum et al. 1991; also see Blum et al. 1992, 1997). The 0–2.5 cm samples were taken because this is the zone where early weed seedling root growth occurs and also because surface residues apparently create an inhibition zone near the surface of the soil which is the location of weed seed germination and early weed seedling growth (DeFrank and Putnam 1978; Putnam et al. 1983). The 0–10 cm samples were taken because this region represents an important portion of the root zone for cover crops, weed seedlings, and crops. Most soil samples collected were extracted in less than a month. After soil samples were defrosted overnight in a refrigerator (10°C), soil subsamples were taken to determine pH, and water, phenolic acid, nitrogen, and carbon content. Soil pH was determined by mixing 25 g soil with 100 ml water, waiting 6 min, and then determining pH with glass electrodes while the slurry was rapidly mixed with a stir bar. Water content was determined gravimetrically on 10–15 g soil subsamples before and after oven drying at 100°C. Nitrogen and carbon content of air-dried soil subsamples were determined with a Perkin-Elmer 2400 C, H, N Analyzer (Perkin-Elmer, Norwalk, Connecticut).

Composition of available/active organic compounds in soil solutions are very complex, being composed of not only phenolic acids but also amino acids, other organic acids, carbohydrates, volatile compounds, etc. (Flaig 1971; Paul and Clark 1989; Lavelle and Spain 2001; Strobel 2001; Blum 2006), any one of these organic compounds at the right concentration and physicochemical state could be inhibitory while at another concentration and physicochemical state they could be neutral or stimulatory to weed seedling germination and emergence (Pandey 1994; Duke at el. 2006; Belz et al. 2007; Belz 2008). Thus weed seedling germination and emergence, as modified by allelopathic agents, is very likely a product of a promoter/organic modifier/inhibitor complex (also other modifiers). The term organic modifiers as used here refers to all neutral organic compounds that indirectly modify the concentrations of promoters or inhibitors by speeding up or slowing down their microbial utilization. For example the preferential utilization of glucose over *p*-coumaric acid by soil microbes (see Sections 2.4.5 and 2.4.6; Blum et al. 1993; Pue et al. 1995; Blum 1996). For other modifiers see Section 2.6.1. The nature of this complex will be determined by the individual organic compounds present, the concentrations of the individual organic compounds making up the complex, their physicochemical states, and their interactions. Inhibition requires that inhibitors dominate in the promoter/organic modifier/inhibitor complex. The focus on phenolic acid is thus somewhat precarious and assumes, rightfully or wrongfully, that phenolic acids are the dominant inhibitory components in the promoter/organic modifier/inhibitor complex, or that phenolic acids are a proxy for all the inhibitors within the promoter/organic modifier/inhibitor complex.

To determine soil phenolic acid content, 50–75 g soil subsamples were placed in 500 ml Erlenmeyer flasks with 100 ml deionized water. Loosely capped flasks were autoclaved (liquid cycle) for 45 min at 1.2 kg/cm^2 and 121°C, removed from the autoclave, and allowed to come to room temperature (20–30 min). The slurry in the flask was then centrifuged for 10 min at 27,200 g. The resulting supernatant was filtered (Whatman No. 42). The filtered supernatant (extract 1) was adjusted to pH 2 with HCL to precipitate humic acids, centrifuged for 10 min at 27,200 g, and the resulting supernatant was adjusted to pH 7 with NaOH (extract 2). Individual phenolic acids based on High Performance Liquid Chromatographic (HPLC) analysis were highly correlated for extract 1 and 2, and total available phenolic acid content (hereafter just total phenolic acid) based on the Folin & Ciocalteu's phenol reagent was proportional to the sum of 7 individual simple phenolic acids quantified. Estimates of total phenolic acid content in extract 1 were only 1.11 times greater than in extract 2. This small difference suggested that little humic acid was recovered by the water-autoclave procedure, thus soil phenolic acid concentrations presented are based on extract 1. Total phenolic acid in soils was determined on water-autoclave extracts and not neutral EDTA (ethylenediaminetetraacetic acid) extracts because EDTA reduces Folin & Ciocalteu's phenol reagent on contact. Additional justifications for the use of the water-autoclave procedure are presented in the Results and Discussion (Section 3.4.1.1). In addition to individual and/or total phenolic acid content, the following were also determined on the water-autoclave soil extracts: pH (glass electrode), freezing point depression (Osmette Precision Osmometer, Precision Systems Inc. Natick, Massachusetts), nitrate and ammonium (Cataldo et al. 1975), phosphorus (Taussky and Shorr 1953), and potassium (atomic absorption spectroscopy). Some soil subsamples were also extracted with water and/or neutral EDTA (see Section 2.2.9). Note: Blum (1997) concluded that neutral citrate extractions could be substituted for neutral EDTA extractions to determine total phenolic acid content of soils. However, Ohno and First (1998) observed that large quantities of organic matter extracted by citrate can mask absorbance due to any extracted simple phenolic acids and thus concluded that the Folin & Ciocalteu's phenol reagent method is not suitable or at best questionable for use with citrate extracts of soils. Citrate extractions and HPLC analysis do, however, provide reliable estimates of individual phenolic acids in soils (Blum 1997).

Phenolic acids in soil extracts were analyzed by HPLC and total phenolic acid content was determined by the Folin & Ciocalteu's phenol reagent (Sigma Chemical Company, St. Louis, Missouri). For HPLC analysis (also see Section 2.2.10), extracts were analyzed using two mobile phases: (A) 2% methanol, 0.25% ethyl acetate, 0.5% acetic acid and 97.25% water, and (B) 80% methanol, 1% ethyl acetate, 2% acetic acid, and 17% water. Linear gradients starting with 92% A and ending with 66% A were used over the first 40 min of a 60-min run. Protocatechuic acid was used as a marker/standard in representative samples. Total phenolic acid content of extracts was determined by the Folin & Ciocalteu's phenol reagent – 0.5 ml extract plus 4.5 ml deionized water or 2.5 ml of a standard phenolic acid solution plus 2.5 ml deionized water were mixed with 0.75 ml of 1.9 M Na_2CO_3 and 0.25 ml Folin & Ciocalteu's phenol reagent (Sigma Chemical Company, St. Louis, Missouri). Mixtures were allowed to stand in the dark at room temperature for at least 1 h before absorption was read at 750 nm (McAllister 1969; Box 1983). Ferulic acid (in some cases other phenolic acids) was used as the standard and thus total phenolic acid values are in ferulic acid (or other phenolic acid) equivalents per gram of soil. Ferulic acid is somewhat intermediate in color development when compared to other phenolic acids (Blum et al. 1991; Blum 2007).

Plant tissues/residues of wheat (*Triticum aestivum* L. "Coker 916 or 983" or "Southern States 555"), rye (*Secale cereale* L. "Abruzzi"), crimson clover (*Trifolium incarnatum* L. "Tibbee"), and/or subterranean clover (*T. subterranean* L. "Mount Barker") were collected from litter bags half buried (i.e., the lower half of the bag was located within the soil) in the field plots (Blum et al. 1991) or from the soil surface (Blum 1997; Lehman and Blum 1997; Staman et al. 2001), freeze-dried, and stored in the dark at room temperature. The freeze-dried plant tissues/residues were ground just before analysis in a Wiley mill (20, 40 or 60 mesh screen) and then extracted and analyzed by several different procedures:

a. Ground subsamples were extracted with water, EDTA, and/or citrate for determination of individual phenolic acids and/or total phenolic acid (Blum 1997). Individual phenolic acids in tissue/residue extracts were analyzed by HPLC (see soil extract analysis above for details, also see Section 2.2.10). Total phenolic acid content of tissue/residue extracts was determined by the Folin & Ciocalteu's phenol reagent (Sigma Chemical Company, St. Louis, Missouri; see soil extract analysis above for details). Total phenolic acid was expressed as ferulic acid (or other phenolic acid) equivalence. Note: Both EDTA and citrate are not very effective in extracting fixed or irreversibly sorbed phenolic acids from plant tissues/residues and EDTA extracts cannot be used to determine total phenolic acid content.
b. Ground subsamples plus water were autoclaved (liquid cycle, 45 min, 1.2 kg/cm^2, and 121°C), allowed to cool, and filtered (Whatman No. 1, Buchner funnel) to produce an extract for determination of individual phenolic acids and/or total phenolic acid (Blum et al. 1991; Lehman and Blum 1997). Individual phenolic acids and total phenolic acid were/was analyzed as in a. and/or
c. Ground subsamples were extracted with 10% methanol in soxhlet extractors to determine total phenolic acid content of tissues/residues by the polyvinylporrolidone (PVP) method (Anderson and Todd 1968; Hall et al. 1982, 1983; Lehman and Blum 1997). Total phenolic acid was expressed as ferulic acid (or other phenolic acid) equivalence.

3.2.2 Laboratory Bioassays

Concentration-dependent inhibition or stimulation of germination and seedling emergence and growth by soil extracts or soil incorporated plant tissues/residues by themselves do not provide irrefutable evidence that allelopathic agents are the causative agents, because one can never be certain whether the observed effects are actually due to the action of allelopathic agents or due to changes in the physicochemical (e.g., pH, water potential, and aeration) and/or biotic (e.g., microbes) environments associated with the testing of soil extracts or plant tissues/residues

(Qasem and Hill 1989; Blum 1999). Bioassay experiments were thus designed to maximize the potential of identifying causative agents by including appropriate controls and including both physicochemical and biotic components in the final statistical/regression analyses (Blum et al. 1992; Lehman and Blum 1997; Staman et al. 2001). For a discussion of the implications and cons and pros of this approach see Blum (1999).

Effects of simulated soil extracts, soil extracts, or serial dilutions of soil extracts on germination, and radicle/epicotyl/hypocotyl length were determined for crimson clover (*Trifolium incarnatum* L. "Tibbee") and/or ivy-leaved morningglory (*Ipomoea heteracea* L.) in petri dishes containing 9-cm-diameter filter paper (Whatman No. 1) and 5 ml of solution (Blum et al. 1992). From now on radicle/epicotyl/hypocotyl lengths will be referred to as radicle/hypocotyl lengths since epicotyls were either absent or very small. Petri dishes were incubated in the dark at 30°C for 48 h before germination and length of radicles/hypocotyls were determined.

Soil samples (0–2.5 cm) for soil extracts were obtained from soybean (*Glycine max* L. "Delapine 417") plots with (not tilled or tilled under) or without wheat stubble and wheat roots (*Triticum aestivum* L. "Coker 916"). Soil samples were extracted by the water-autoclave procedure (Section 3.2.1). Regression analyses were used to determine the relationship between simulated soil extract or initial (undiluted) soil extract characteristics (e.g., pH, phenolic acid content, freezing point depression, NO_3^-, NH_4^+, K, and/or P; see Section 3.2.1) and seed germination, radicle/hypocotyl lengths, or the slopes for radicle/hypocotyl lengths obtained from serial dilutions of soil extracts. The range of solute potentials for simulated soil extracts were created with polyethylene glycol 6000 or different strength of Hoagland's solution (Blum et al. 1992). The solute potentials of simulated soil extracts and undiluted soil extracts were estimated by relating freezing point depressions of extracts with solute potentials based on freezing point depression of sodium chloride (Lang 1967) or a model for polyethylene glycol 6000 (Steuter 1981; Blum et al. 1992). Seed and seedling response to extracts, however, were based on freezing point depression values.

Effects of cover crop tissues/residues on seedling emergence of red-root pigweed (*Amaranthus retroflexus* L.) and morningglory (*Ipomoea hederacea* L.) were determined in petri dishes containing different mixtures of soil (1 unit of soil:2 units of sand by weight) and cover crop tissues/residues (Lehman and Blum 1997; Staman et al. 2001). Cover crop tissues/residues of wheat (*Triticum aestivum* L. "Coker 983" or "Southern States 555"), crimson clover (*Trifolium incarnatum* L. "Tibbee"), rye (*Secale cereale* L. "Abruzzi"), or subterranean clover (*T. subterranean* L. "Mount Barker") were collected before and/or various times after glyphosate desiccation in the field, frozen, freeze dried, and stored with desiccant in the dark at room temperature. Prior to each experiment, freeze-dried tissues/residues were ground in a Wiley mill (20 or 40 mesh screen) and incorporated at various amounts into 150 g Cecil A soil-sand mixture (1 soil:2 sand by weight). Twenty five morningglory or 50 pigweed seeds were placed into the soil-tissue/residue mixture just below the soil surface in loosely covered petri dishes (100 × 25 mm) and placed in a growth chamber with a 12 h photoperiod and a photosynthetic photon flux density (PPFD) of 411 μmol/m^2/s (Lehman and Blum 1997) or under a light bank with a 12-h photoperiod and PPFD of 140 μmol/m^2/s (Staman et al. 2001). At 24-h intervals seedling emergence was recorded and seedling removed. Deionized water was added daily by weight to bring soil moisture back to the original treatment levels. Experiments were terminated at 4 (morningglory) or 6 days (pigweed). Effects and interactions of temperatures (25/21, 30/26, and 35/31°C), moisture (0.093–0.173 g water/g soil), nitrogen (0–14 μg N/g soil), and cover crop tissues/residues (0–15 mg/g soil) on seedling emergence were also determined (Lehman and Blum 1997).

Effects of phenolic acids, wheat (*Triticum aestivum* L. "Southern States 555") or sunflower tissues (*Helianthus annuus* L.) on phenolic acid-utilizing bacteria in the bulk soil and rhizosphere of cucumber seedlings and growth inhibition of cucumber seedlings were determined (Staman et al. 2001) as previously described. For details see Section 2.2.11. The effect of sunflower leaf tissue on pigweed seedling emergence was also determined, as described above.

3.2.3 Field Studies

3.2.3.1 Field Weed Seedling Emergence

In the fall experimental areas with Cecil A soils (Typic Hapludults, clayey, kaolinitic, thermic, also see Table 2.1) were disked and wheat (*Triticum aestivum* L. "Coker 983" or "Southern States 555"), rye (*Secale cereale* L. "Abruzzi"), crimson clover (*Trifolium incarnatum* L. "Tibbee"), or subterranean clover (*T. subterranean* L. "Mount Barker") were planted in 24 m^2 plots (Blum et al. 1997, 2002). The cover crops were desiccated with glyphosate (N-[phosphonomethyl] glycine) at various times during April and/or May (no-till plots) of 1992, 1993, 1996, and 1997 or in May of 1993 living cover crops were cut with a flail mower and subsequently incorporated into the soil with a rotary tiller (tilled plots). Plots were desiccated at different times to determine the effects of the timing of glyphosate desiccation on weed seedling emergence. No-till reference plots without cover crops for desiccation treatments and tilled reference plots without cover crops for incorporated treatments were included. In addition desiccated cover crops were cut near the soil surface and left in place or moved to plots without cover crops in 1996 and 1997 (Blum et al. 2002).

Weed seeds (ivy leaf morningglory, *Ipomoea hederacea* L.; prickly sida, *Sida spinosa* L.; red-root pigweed, *Amaranthus retroflexus* L.) purchased from V&J Farms (Woodstock, Illinois) or Azlin Seed Service (Leland, Mississippi) were broadcast onto untilled (no-till) or tilled (only in 1993) surfaces of 1.86 m^2 alternating subplots (one subplot per weed species; 300–2,500/m^2 depending on species or year) in all 24 m^2 experimental plots (Fig. 3.1). Alternating subplots were chosen at random for each species. The two end subplots were excluded. Weed seeds were broadcast onto no-till surfaces several months prior to glyphosate desiccation of

Fig. 3.1 Frame used to determine location of subplots for weed seeds. Location of subplot for each weed species within each treatment plot was chosen at random. The two outer subplots were not used

cover crops and onto tilled surfaces just after cover crop incorporation. Emerging weed seedlings were counted in the subplots at various times during May and July. To prevent seed production of weeds in the experimental plots, all vegetation including weed seedlings were killed in the plots prior to flowering.

Surface tissue/residue of cover crops, solar radiation reaching the soil surface, soil temperature, soil compaction, soil moisture, soil pH (1:2 soil:water), soil nitrate-N, and/or phenolic acid content of soil were determined at various time intervals during the experimental periods (Blum et al. 1991, 1992, 1997, 2002). Amounts of cover crop living biomass or surface residues were based on 0.25 m^2 subplots. Solar radiation (blue [400–500 nm] and red [600–700 nm]) reaching the soil surface was determined with a Plant Growth Photometer (International Light Inc., Massachusetts). Soil temperatures were taken at the soil surface and at 2.54 cm with YSI Tele-thermometer probes (Yellow Springs Instrument Co., Yellow Springs, Ohio). Soil compaction was determined with an impact penetrometer (345 g) using a 4- by 65-mm probe that was dropped from 35 cm above the soil surface. Soil moisture (100°C), pH (1:2 soil:water), and C/N ratio (Perkin-Elmer 2400 C, H, N, Analyzer; Perkin-Elmer, Norwalk, Connecticut) were determined on subsamples taken from 0 to 2.54 cm and 0–10 cm soil cores. Soil subsamples from the soil cores were extracted by a water-autoclave procedure (Blum et al. 1991, 1992) and extracts were analyzed for individual phenolic acids (HPLC analyses), total phenolic acid (Folin and Ciocalteu's phenol reagent; Blum et al. 1991, 1992), pH (glass electrode), and nitrate-N (Cataldo et al. 1975). For additional details see Section 3.2.1.

3.2.3.2 Cover Crops

Wheat (*Triticum aestivum* L. "Coker 983" or "Southern States 555"), rye (*Secale cereale* L. "Abruzzi"), crimson clover (*Trifolium incarnatum* L. "Tibbee"), and subterranean clover (*T. subterranean* L. "Mount Barker") were used as cover crops. These cover crops were chosen because of their different physical and chemical characteristics (Fig. 3.2). Crimson and subterranean clovers are composed primarily

a

b

Fig. 3.2 Cover crops before they were desiccated with glyphosate (**a**): crimson clover (*front right*), subterranean clover (*front left*), wheat (*back right*) and rye (*back left*; Blum et al. 1997). Wheat plots after they were desiccated with glyphosate (**b**): shoots cut and uncut and reference plot in the right-hand corner (Blum et al. 2002) of petioles and runners while wheat and rye are composed primarily of leaves and stems. The chemical composition of these species also varied greatly; e.g., the C/N ratio, fiber, and total phenolic acid content were different (Morrison 1956; Li et al. 1992; Lehman and Blum 1997; Stubbs et al. 2009). Since chemical composition of cover crops also varies with plant age, additional differences in chemical composition of cover crop tissues/residues were generated by desiccating the cover crops at different times.

3.2.3.3 Weed Species

Ivy leaf morningglory (*Ipomoea hederacea* L.), prickly sida (*Sida spinosa* L.), and red-root pigweed (*Amaranthus retroflexus* L.) were chosen because they are common aggressive weeds in agricultural systems in the Southeastern United States. Given the size of the experimental plots and the heterogeneity of weed populations in agricultural fields, the effects of cover crops on individual species of weeds are difficult to determine at a single location. Thus we broadcast purchased weed seeds in January or February onto untilled (no-till) surfaces of the subplots (one weed species per subplot). Since weeds (Fig. 3.3) did not emerge until May or June, there was ample time for seed populations to be influenced by predation, mortality, and environmental fluctuations. The tilled subplots were seeded after they were tilled in late April shortening the time of environmental influences on the weed seed population. The use of local weed seeds would have been better but a consistent source and an adequate supply of morningglory, pigweed and prickly sida for these studies was not readily available. To prevent seeded weed populations from producing seeds that could influence results of subsequent growing seasons, the entire study area was treated with glyphosate to kill all living vegetation before weed seed production. Weed seedling emergence was primarily based on the total number of weed seedlings in the subplots over time. The weed seedling numbers observed were thus a product of the number of viable seeds broadcast, weed seeds in the soil bank, proportion of dormant vs. non-dormant (quiescent) seeds, the action of seed predators and pathogens, the environmental conditions for germination, success of conversion from a heterotrophic to an autotrophic state, and subsequent seedling survival. Thus weed seedling populations, as determined here, not only increased but also decreased over time.

3.2.3.4 Agricultural Practices

Although normal agricultural practices were followed where possible, clearly a number of practices (e.g., seeding of weeds, the absence of actual crops [e.g., soybean, corn] in some instances, the timing of cover crop desiccation, and the desiccation of all vegetation in the plots prior to weed seed production) were inconsistent with normal agricultural practices. Since the focus of this research was on the influence of early weed seedling emergence instead of weed management and crop production, these inconsistencies with normal agricultural practices were acceptable, actually essential in meeting some of our experimental objectives.

a

b

Fig. 3.3 Weed seedlings in wheat plots at end of an experimental period: (**a**) morningglory *upper right corner* and prickly sida *center*, and (**b**) pigweed *center* and morningglory *lower left*

3.2.3.5 Physicochemical Environment

Even though considerable time and energy were expended in characterizing the physical and chemical properties of the soil in the field plots, the resulting data obtained were, at best, crude average approximations since local variation from point to point, time of day, and season are highly variable. In addition average physical and chemical environmental data for given locations and points of time are not necessarily representative of what is experienced by individual weed seeds and weed seedling roots in the subplots. Thus the usefulness of such data in identifying cause and effect may be of limited value. They do, however, provide a general picture of potential differences in environments created by the presence/absence of different types of cover crop tissue/residue and management practices.

3.2.4 Data Analyses

Laboratory and/or field data were analyzed using SAS systems (SAS Institute Inc. 1999) utilizing analysis of variance, regression analysis, response surface analysis, univariate analysis, repeated measures analysis (multivariate profile analysis), covariance analysis and/or principle components analysis. Good statistical practices were used to verify that the data satisfied the assumptions underlying the various analyses. Significant differences between means were determined by Tukey's Studentized Range Test, the Tukey-Kramer HSD test, or the Bonferroni *t* test. Alpha was set at 0.05.

3.3 Research Objectives

The primary research objectives were as follows:

a. To characterize phenolic acids in soils of no-till and conventional-till systems and to establish correlations between easily obtained soil characteristics and phenolic acids in soils.
b. To determine if soil extracts could be used directly in laboratory bioassays for the detection of allelopathic activity.
c. To characterize how cover crop residues in no-till systems affect early emergence of broadleaf weeds and to establish and characterize potential relationships between early broadleaf weed seedling emergence and the physical and chemical environments resulting from the presence of cover crop residues.
d. To characterize cover crops and cover crop residues and how these may potentially modify the soil environment.
e. To determine under controlled conditions how effects of shoot cover crop residues taken from the field change with time after desiccation and how such effects are modified by temperature, moisture, and nitrogen levels.
f. To determine the respective importance of shoot and root residues in regulating early broadleaf weed seedling emergence, and
g. Determine under controlled conditions how phenolic acid-containing plant tissues/residues mixed into soil modify phenolic acid-utilizing bulk-soil and rhizosphere microbial populations.

3.4 Results and Discussion

3.4.1 Characterize the Phenolic Acids in Soils of No-Till and Conventional-Till Systems and to Establish Correlations Between Easily Obtained Soil Characteristics and Phenolic Acids in Soils (Blum et al. (1991); Plenum Publishing Corporation, Excerpts Used with Permission of Springer Science and Business Media)

Since phenolic acids released from wheat tissues/residues have been implicated in the inhibition of the emergence of broadleaf weed seedlings (Liebl and Worsham 1983; Shilling et al. 1985; Worsham 1989, 1990; Putnam et al. 1983; Blum et al. 1991, 1997, 2002; Lehman and Blum 1997; Staman et al. 2001), we utilized wheat stubble/soybean (no-till), wheat stubble tilled under/soybean (conventional-till), and fallow/soybean (conventional-till) systems for this study. Both wheat and soybean roots were present in the wheat plots.

Phenolic acids in soils occur either in a "free" state in the soil solution, reversibly sorbed to soil particles, fixed (irreversibly sorbed) very tightly to soil particles (e.g., recalcitrant organic matter, and clays), and/or on and in living and dead plant tissues/residues ("free", reversibly sorbed, and fixed). Of general interest to plant–plant allelopathic interactions are the "free" and reversibly sorbed states frequently referred to as the available fraction. Of particular interest is the active fraction of available phenolic acids, the fraction of available phenolic acids that actually interact with seeds, roots and microbes. Unfortunately we presently do not have a means of quantifying the active fraction, thus the focus on the available fraction.

Phenolic acids in soils range from simple compounds such as benzoic acid derivatives to complex polymers such as tannins (Figs. 3.4 and 3.5). There is also a large array of polymers that contain phenolic acid moieties (e.g., lignin, humic acids,

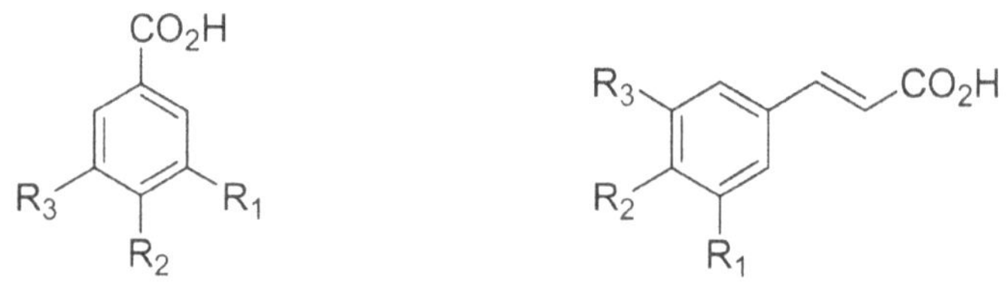

R_1	R_2	R_3	
H	OH	H	*p*-hydroxybenzoic acid
OH	OH	H	protocatechuic acid
OMe	OH	H	vanillic acid
OMe	OH	OMe	syringic acid

R_1	R_2	R_3	
H	OH	H	*p*-coumaric acid
OH	OH	H	caffeic acid
OMe	OH	H	ferulic acid
OMe	OH	OMe	sinapic acid

Fig. 3.4 Some common simple plant phenolic acids, cinnamic acid derivatives on the *right* and benzoic acid derivatives on the *left*, where H equals hydrogen, OH equals hydroxy, and OMe equals methoxy

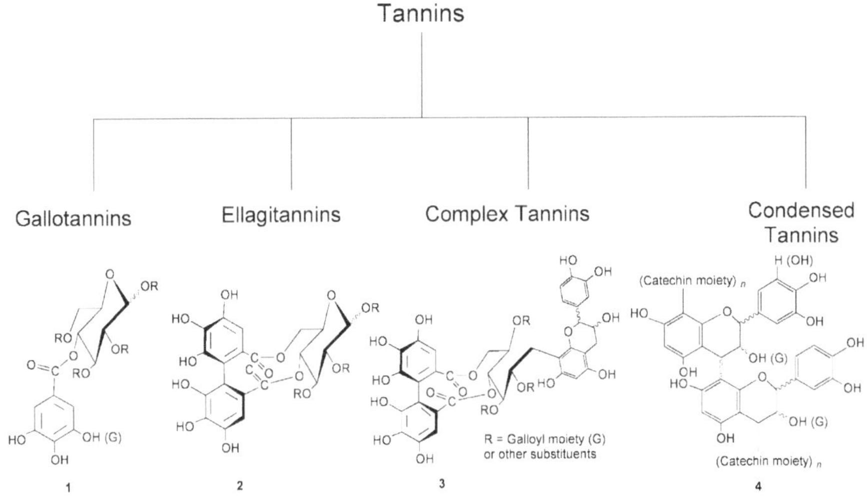

Fig. 3.5 Tannins. Figure reproduced from Khanbabaee and van Ree (2001). Figure used with permission of the Royal Society of Chemistry

fulvic acids, and melanins; Fig. 3.6). To determine concentrations of phenolic acids in soils ("free", reversible sorbed, or fixed; simple to complex polymers) requires some type of extraction or recovery procedure, none of which are 100% efficient, and some type of isolation and quantification technique which varies with the type of phenolic acid to be isolated and quantified (Hartley and Whitehead 1985; Dalton et al. 1987; Blum et al. 1991; Harborne 1998; Vermerris and Nicholson 2006). Thus absolute total values for phenolic acid concentrations in soils are impossible to determine. Here of course we are interested in the available phenolic acids, but even absolute values for total available phenolic acids are difficult, if not impossible, to determine (see below). However, reasonably accurate estimates of available fractions of individual phenolic acids in soils, such as benzoic and cinnamic acid derivatives, are achievable.

Since water is the primary carrier and transport medium in the soil, some type of water extractant was desirable. Thus most researchers would agree that extractions by water, low concentrations of calcium hydroxide, sodium acetate, or mild chelating agents provide the most biologically meaningful estimates for the available fraction of phenolic acids in soils. We have found for our soils that reasonable estimates of available soil concentrations for individual simple phenolic acids were obtained by neutral EDTA or citrate extractions and isolation and quantification by HPLC analysis (Blum et al. 1994; Blum 1997; also see Section 2.4.3). Unfortunately EDTA reduces the Folin & Ciocalteu's phenol reagent on contact so that available total phenolic acid content of soils could not be determined for EDTA extracts. Citrate extracts do not reduce the Folin & Ciocalteu's phenol reagent but should only be used to estimate total phenolic acid content of soil with extremely low organic content (Ohno and First 1998). Since we wished to get estimates of concentrations

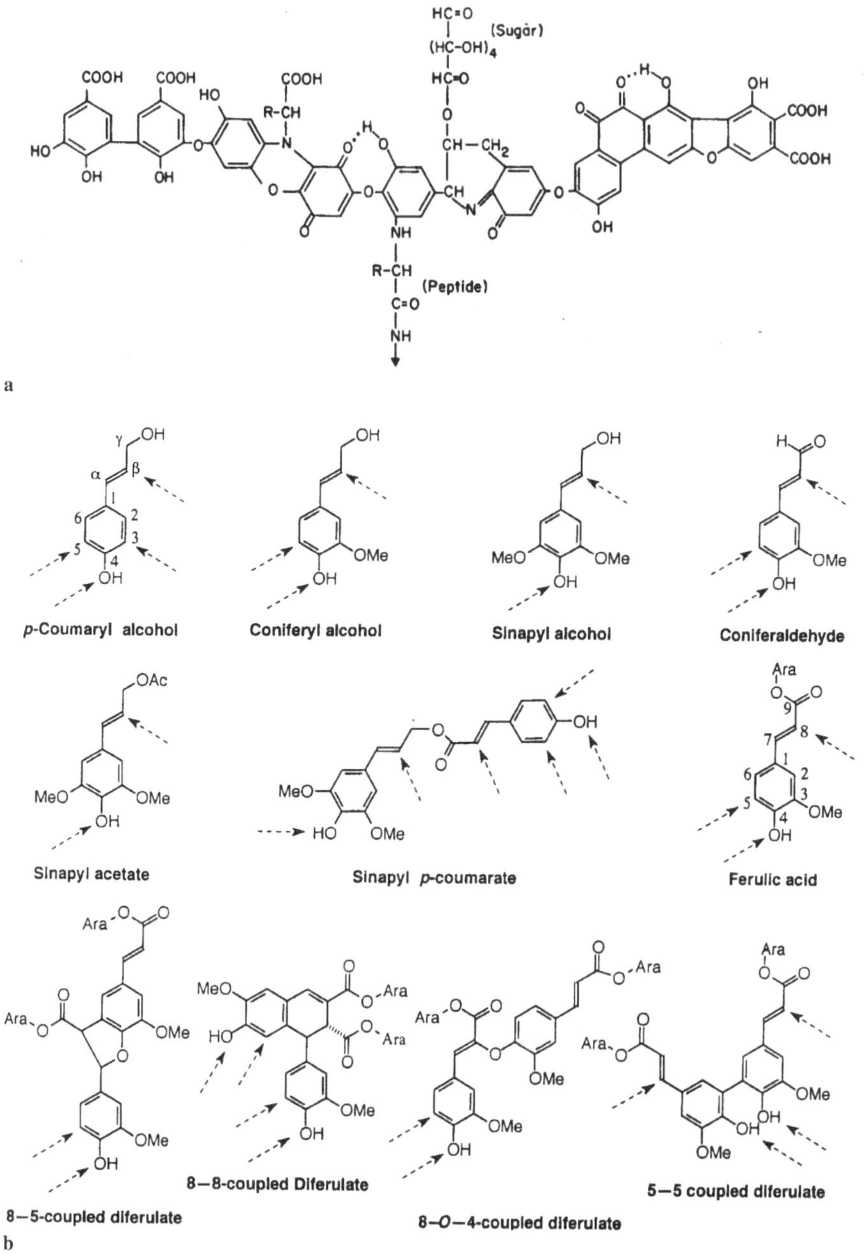

Fig. 3.6 Two examples of polymers that contain phenolic acid moieties: (**a**) model of humic acid and (**b**) precursors of lignin. Figure (**a**) reproduced from Stevenson (1982) and (**b**) from Grabber (2005). Figure (**a**) and (**b**) used with permission of John Wiley and Sons, Inc and Crop Science Society of America, respectively

for both available individual and total phenolic acid in soils, we utilized another extraction procedure for this study, a water-autoclave procedure (Blum et al. 1991). The reasons for identifying both individual and total phenolic acids in soils were as follows:

a. The potential inhibitory activity of individual phenolic acids on seedlings varies considerably. For example cinnamic acids are frequently more inhibitory than benzoic acid derivatives in nutrient culture and actual observed inhibition of individual phenolic acids in soil systems varies with soil type, pH, etc. (Blum et al. 1985a, b, 1989; Gerig and Blum 1991; Blum and Gerig 2006).
b. For a given percent inhibition, the required concentrations of individual phenolic acids in a mixture decline as the number of phenolic acids in a mixture increased (Gerig and Blum 1991; Blum 1996).
c. The inhibitory activity of individual phenolic acids in a mixture was additive at lower concentrations and partially antagonistic at higher concentrations and the concentration thresholds for additive and antagonistic effects varied with phenolic acid (Blum 1996), and
d. Soils contain a complex array of phenolic acids, including polymers that contain phenolic acid moieties, many of which may be involved in plant–plant allelopathic interactions.

Thus the summed or average action of extracted available total phenolic acids in soils is just as important as understanding the action of available individual phenolic acids within such extracts.

There is, however, a caveat for estimating available total phenolic acid concentrations. The estimates of the total available fraction of phenolic acids in soil extracts represent a crude estimate of what actually occurs in soil, not only because of the range of efficiencies of extraction procedures but also because different phenolic acids at the same concentration generate different absorbances with Folin & Ciocalteu's phenol reagent (Fig. 3.7; Blum et al. 1991). In addition soil extracts also contain compounds, other than phenolic acids, that react with (i.e., reduce) the Folin & Ciocalteu's phenol reagent (McAllister 1969; Box 1983). The assumption, therefore, was that available total phenolic acid values based on the Folin & Ciocalteu's phenol reagent expressed as ferulic acid equivalence were relative values that were consistently related to the actual total available phenolic acids (hereafter just called total phenolic acid) present in soil extracts. The extraction and quantification by HPLC analysis of available individual phenolic acids in soil do not have these particular problems.

However, there is one more caveat for both total and individual phenolic acids. Soil extractions recover residual or net concentration, i.e., input – losses, for a point or various points in time. Since both input and losses are unknown between points of time the actual available total or individual phenolic acid concentrations in soil over time are also unknown. The concentrations of available phenolic acids interacting with roots could thus be greater, at times much greater, or lower, at times much lower, than the net concentrations determined from soil extracts.

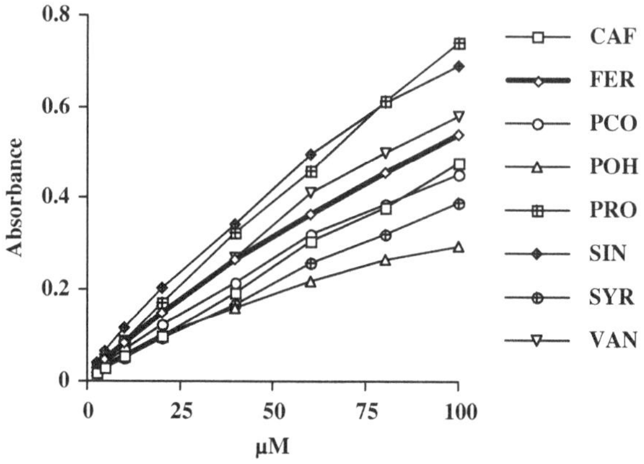

Fig. 3.7 Standard curves. Absorbance of caffeic acid (CAF), ferulic acid (FER), *p*-coumaric acid (PCO), *p*-hydroxybenzoic acid (POH), protocatechuic acid (PRO), sinapic acid (SIN), syringic acid (SYR), and vanillic acid (VAN) after reacting with the Folin & Ciocalteu's phenol reagent. Figure reproduced from Blum et al. (1991). Plenum Publishing Corporation, figure used with permission of Springer Science and Business Media

3.4.1.1 Water-Autoclave Procedure

Simple phenolic acids, such as ferulic, *p*-coumaric, vanillic, etc., are slow to dissolve in water at ambient temperatures. The speed at which phenolic acid dissolve and the amount of phenolic acid that dissolves in water, however, can be increased by increasing the pH and/or the temperature of water. Increasing the temperature and/or pH of a water extraction solution may at times be more desirable than using long-term ambient water extractions because of the action of microbes. Blum and Shafer (1988) for example noted that $< 1\%$ of exogenously applied ferulic, vanillic, *p*-coumaric or *p*-hydroxybenzoic acids could be extracted with water 24 h after addition of phenolic acids to non-sterilized Portsmouth A soil samples but $\geq 95\%$ could be recovered from sterilized Portsmouth A soil samples. Thus extracting soils for extended periods to recover available forms of phenolic acids could lead to substantial underestimates while artificially high pH or temperatures could lead to overestimates of available phenolic acids present in soil solutions. The water-autoclave extraction was thus a compromise. It is a rapid (high temperature and pressure) procedure, has only a minimal impact on pH, and eliminates the concern about microbial activity. With the exception of caffeic acid (approximately 20% reduction) and sinapic acid (approximately a 61% reduction) no major losses or gains were observed when distilled water containing mixtures of phenolic acids composed of caffeic, ferulic, *p*-coumaric, *p*-hydroxybenzoic, protocatechuic, sinapic, syringic, and vanillic acids were autoclaved (Blum et al. 1991). Soil phenolic acid values presented here for the water-autoclave extracts have not been adjusted for potential reductions in caffeic acid and sinapic acid because phenolic acid losses in soil solutions are likely to be different from losses in distilled water solutions.

Water-autoclave extractions were also much more efficient than neutral EDTA extractions in recovering sorbed/fixed phenolic acids from plant tissues/residues (Blum et al. 1992). For example HPLC analysis of EDTA extracts for wheat stubble collected after a wheat harvest contained no detectable phenolic acid peaks. Water-autoclave extracts of this wheat stubble had 11 distinct peaks. Concentrations for ferulic acid, vanillic acid, *p*-coumaric acid, and *p*-hydroxybenzoic acid were 33, 22, 1034, and 47 μg/g dry weight, respectively (Fig. 3.8). On the other hand water and

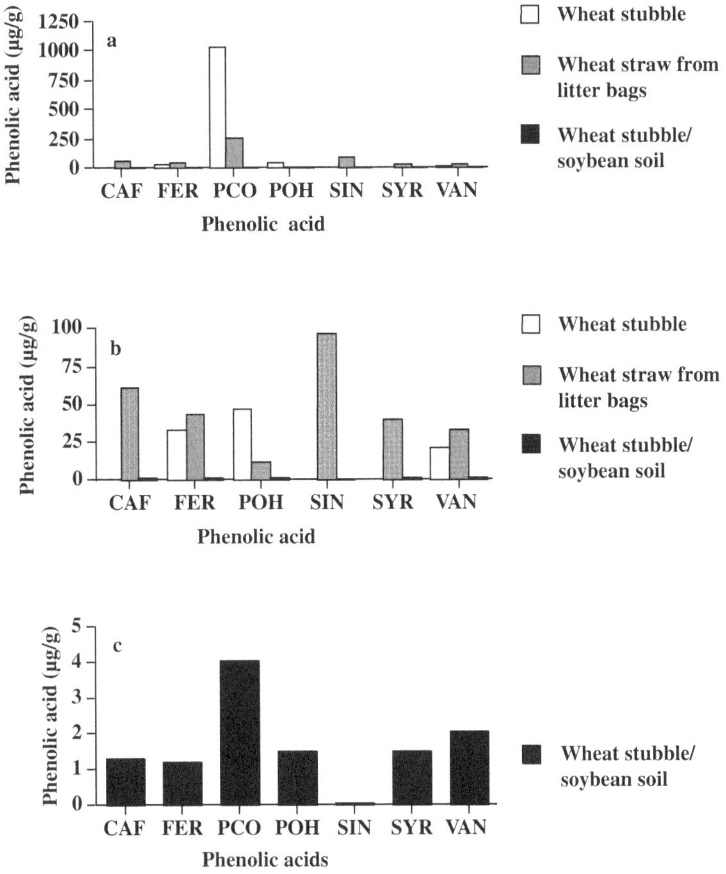

Fig. 3.8 Phenolic acids extracted from wheat stubble, wheat straw from half buried litter bags, and wheat stubble/soybean (no-till) soil. Phenolic acids isolated and quantified were caffeic acid (CAF), ferulic acid (FER), *p*-coumaric acid (PCO), *p*-hydroxybenzoic acid (POH), sinapic acid (SIN), syringic acid (SYR), and vanillic acid (VAN). Because *p*-coumaric acid was so high in comparison to other phenolic acids in wheat residues, data are presented twice, once with *p*-coumaric acid (**a**) and once without *p*-coumaric acid (**b**). Because phenolic acids were so low in the soil they are also presented in (**c**). The absence of standard error bars for wheat straw and soil indicates that the error bars are too *small* to be visible. Figures based on data from Blum et al. (1991, 1992). Plenum Publishing Corporation, data used with permission of Springer Science and Business Media

neutral EDTA extractions in conjunction with HPLC analysis clearly provided very reliable data on the biologically available fractions ("free" and reversibly sorbed) of simple individual phenolic acids in phenolic acid amended Cecil soil samples (see Section 2.4.3; Blum et al. 1992, 1994). The question at this point: How similar or different are the estimates of available phenolic acids in soils provided by the neutral EDTA and the water-autoclave procedures?

One hundred days after the addition of 1,000 μg/g (approximately 5 μmol/g soil) ferulic acid to sterile Cecil soil samples, water, neutral EDTA, and water-autoclave extractions recovered 28% (277 μg/g), 37% (373 μg/g), and 32% (322 μg/g) of the ferulic acid added, respectively, from Cecil A soil samples (3.7% organic matter) and 17% (167 μg/g), 52% (524 μg/g), and 30% (304 μg/g) of the ferulic acid added, respectively, from Cecil B soil samples (0.2% organic matter; Blum et al. 1992). The recovery of ferulic acid from soil by neutral EDTA extraction was thus more effective than the water-autoclave extraction.

For soil samples that had been inoculated with microorganisms 70 days after the addition of ferulic acid, recovery was reduced to less than 2% 30 days after soil samples were inoculated (for Cecil A soil samples 0, 4, and 20 μg/g were recovered, respectively; for Cecil B soil samples 0, 0.4, and 4 μg/g were recovered, respectively). More of the ferulic acid recovered by neutral EDTA was available to soil microbes. This indicated that there was not only a quantitative difference but also a qualitative difference in the fractions of sorbed ferulic acid recovered by the two extraction procedures. By fractions I am referring to ferulic acids bound by different kinds and strengths of bonds to soil particles (see Section 2.4.3.2).

We suspect that both neutral EDTA and water-autoclave extraction procedures readily recovered the "free" phenolic acid fraction. So how effective were they in extracting reversibly sorbed phenolic acids (i.e., sorbed recovered by neutral EDTA or water-autoclave extraction minus "free" recovered by water extraction)? For the 1,000 μg/g added 723 μg/g was sorbed by Cecil A and 833 μg/g by Cecil B. For Cecil A samples, 13% (96 of 723 μg/g) and 6% (45 of 723 μg/g) of the sorbed ferulic acid was recovered by the neutral EDTA and the water-autoclave procedures, respectively. Of this 96% (92 μg/g) and 55% (25 μg/g), respectively, was utilized by microorganisms over the 30 days (Blum et al. 1992). For Cecil B samples, 43% (357 of 833 μg/g) and 16% (137 of 833 μg/g) of the sorbed ferulic acid was recovered by the neutral EDTA and the water-autoclave procedures, respectively. Of this 100% (357 μg/g) and 97% (133 μg/g), respectively, was utilized by microorganisms over the 30 days.

In conclusion, the water-autoclave extraction procedure when compared to the EDTA extraction procedure underestimated the total available ferulic acid in the soil by roughly 5% for Cecil A and 22% for Cecil B. In addition to the quantitative difference there also appeared to be a difference in the types of the sorbed ferulic acid recovered. The water-autoclave-procedure recovered some irreversibly sorbed phenolic acids from Cecil A soil since only 55% of the sorbed phenolic acid recovered was utilized by microbes. This difference should not be surprising since the physical and chemical processes of the two extraction procedures, i.e., chelation vs.

temperature and pressure, are different. For our field studies Cecil A was the primary soil type.

3.4.1.2 Soil Phenolic Acids

So what were the concentrations of phenolic acids in the Cecil A soil ± wheat stubble (*Triticum aestivum* L. "Coker 916")/soybean (*Glycine max L.* "Deltapine 417") systems? Subsamples taken from wheat stubble/soybean (no-till), wheat stubble tilled under/soybean (conventional-till), and fallow/soybean soil (conventional-till) cores were extracted by the water-autoclave procedure and analyzed for 7 common phenolic acids (ferulic, caffeic, *p*-coumaric, p-hydroxybenzoic, sinapic, syringinc, and vanillic) and total phenolic acid (Blum et al. 1991). With minor exception, individual phenolic acids were correlated with each other, with the sum of the 7 phenolic acids identified by HPLC analysis, and total phenolic acid as determined by the Folin & Ciocalteu's phenol reagent method.

Average individual phenolic acids for the wheat stubble/soybean 0–2.5 cm soil samples for the 109-day-experimental period were as follows: *p*-coumaric acid 4.08 ± 0.22, vanillic acid 2.06 ± 0.11, syringic acid 1.53 ± 0.07, p-hydroxybenzoic acid 1.51 ± 0.06, caffeic acid 1.30 ± 0.06, ferulic acid 1.21 ± 0.08, sinapic acid 0.06 ± 0.07, and total (sum of 7) 12.30 ± 0.58 μg/g soil ± standard error (Figs. 3.8 and 3.9). Similar patterns, but lower concentrations, were observed for wheat stubble tilled under/soybean and fallow/soybean soil samples (Fig. 3.9). In

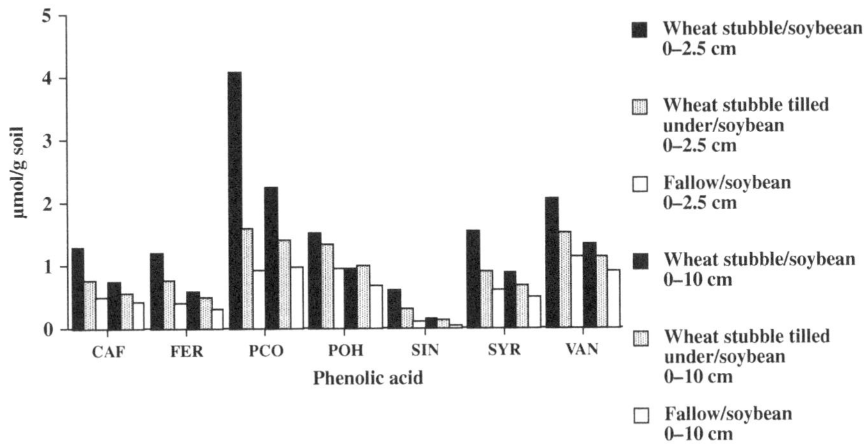

Fig. 3.9 Phenolic acids extracted from wheat stubble/soybean (no-till), wheat stubble tilled under/soybean (conventional-till), and fallow/soybean (conventional-till) Cecil A soils for 0–2.5 and 0–10 cm soil cores. Phenolic acids isolated and identified were caffeic acid (CAF), ferulic acid (FER), *p*-coumaric acid (PCO), *p*-hydroxybenzoic acid (POH), sinapic acid (SIN), syringic acid (SYR), and vanillic acid (VAN). The absence of standard error bars indicates that the error bars are too *small* to be visible. Figure based on data from Blum et al. (1991). Plenum Publishing Corporation, data used with permission of Springer Science and Business Media

general the benzoic acids (vanillic, *p*-hydroxybenzoic and syringic) were higher in concentration than the cinnamic acids (caffeic, ferulic, and sinapic), *p*-coumaric acid being the exception. The highest concentration observed for any phenolic acid was 4.08 ± 0.22 μg/g soil for *p*-coumaric acid. The range of values for individual phenolic acids obtained here were similar to what others have found utilizing a variety of other mild extraction procedures for other soil systems (Whitehead et al. 1981; Hartley and Whitehead 1985; Siqueira et al. 1991). The pattern for individual phenolic acids from the 0–10 cm soil samples were similar to those from 0 to 2.5 cm samples except that the concentrations were lower (Fig. 3.9). The sum of the 7 phenolic acids for wheat stubble/soybean, wheat stubble tilled under/soybean, and fallow/soybean soil samples recovered from the 0–2.5 cm soil cores were 12.30 ± 0.58, 7.18 ± 0.32, and 4.64 ± 0.34 μg/g soil, respectively. For the 0–10 cm soil cores these values were 6.89 ± 0.38, 5.83 ± 0.14, and 3.83 ± 0.13 μg/g soil, respectively. The higher amounts in the 0–2.5 cm cores compared to the 0–10 cm cores were associated with the higher average organic matter (e.g., litter residue and recalcitrant organic matter) of the 0–2.5 cm core samples. Soil total phenolic acid (ferulic acid equivalents) based on the Folin & Ciocalteu's phenol reagent method was, on average, 15 times greater than the sum of the 7 phenolic acids.

Whitehead et al. (1981) suggested that substantial portions of phenolic acids extracted from soils were either derived from organic residues that were more than 4 years old or were the result of microbial activity. Our findings (Blum et al. 1991) appear to be consistent with their observations since:

a. The concentrations of individual phenolic acids, the sum of the 7 phenolic acids, and total phenolic acid extracted from soil samples were not only significantly related to each other but also to soil pH, water content of soil samples, total soil carbon, and total soil nitrogen.
b. No dramatic changes or clear trends for phenolic acid concentrations in soil samples were observed over the first 109 days of the growing season. Maximum changes for the sum of 7 phenolic acids for the different sampling dates were on the order of 20–30%. However, the actual phenolic acid turnover rates are not known since real time source-sink relationships in the soil cannot be determined from soil extracts at 2–4 week intervals, and
c. Total phenolic acid content based on the sum of 7 phenolic acids of incubated (dark, 30°C) soil samples from 0 to 2.5 cm soil cores treated with water or nutrient solution declined by 8% and increased by 3%, respectively at the end of the first week and declined by13% and increased by 18% at the end of the second week, respectively (Blum et al. 1991). For 0–10 cm soil samples values decreased 0 and 10% at the end of week one, respectively.

No additional changes occurred over the next week. These patterns of decline and/or increase over a 2-week period suggested not only that phenolic acids in the soil samples were primarily in a bound (sorbed) form but also that microbial utilization/synthesis of available phenolic acids was nutrient limited.

3.4.1.3 Summary

Mean concentrations of available individual benzoic and cinnamic acid derivatives determined in this Cecil soil were small, less than 4 μg/g soil. The sums of 7 individual phenolic acids (0–2.5 cm soil cores) for wheat stubble tilled under/soybean and fallow/soybean soil samples were 58 and 38%, respectively, of wheat stubble/soybean soil samples (100% =12.30 ± 0.58 μg/g). The sum of 7 individual phenolic acids for the 0–2.5 cm core samples was approximately 34% higher than for the 0–10 cm core samples. Plant tissues/residues contained greater individual phenolic acid content than soils. For example wheat stubble contained 258 times and wheat straw from half buried litter bags 65 times the *p*-coumaric acid of wheat no-till Cecil A soils (4 μg/g soil).

Water-autoclave extractions of soils in conjunction with HPLC analysis appeared to provide reasonable estimates of available individual phenolic acids in soil. The majority of the individual phenolic acids recovered by the water-autoclave extraction procedure appeared to be released from sorbed (fixed) forms. Thus the primary source of available individual phenolic acids in these soils during most of the growing season would very likely be associated with the action of microbes on plant tissues/residues and/or soil organic matter. Since soils contain numerous active phenolic acid type compounds other than benzoic and cinnamic acid derivatives we also estimated available total phenolic acid content of soils for the water-autoclave soil extracts. Soil total phenolic acid (ferulic acid equivalents) based on the Folin & Ciocalteu's phenol reagent method was, on average, 15 times greater than the sum of the 7 phenolic acids. However, estimates of soil total phenolic acid based on water-autoclave extraction and Folin & Ciocalteu's phenol reagent are likely overestimated since other reactive compounds besides phenolic acids are included, this in spite of the fact that the water-autoclave procedure recovered lower available phenolic acid concentrations from phenolic acid "spiked" soils than the neutral EDTA procedure, our standard (baseline) extraction procedure. Finally, the concentrations of individual, sum of 7 phenolic acids, and total phenolic acids recovered by the water-autoclave extraction procedure were significantly and linearly related to each other, soil pH, soil organic matter, total soil carbon, and total soil nitrogen.

3.4.2 Determine if Soil Extracts could be Used Directly in Laboratory Bioassays for the Detection of Allelopathic Activity (Blum et al. (1992); Plenum Publishing Corporation, Excerpts Used with Permission of Springer Science and Business Media)

Germination and radicle/hypocotyl growth of weed seeds are modified by many physical and chemical characteristics of soil extracts such as presence, concentrations, and activities of promoters and inhibiters, pH, solute potential, and organic and inorganic compounds. Effects of soil extracts on seed germination and radicle/hypocotyl growth are thus not the exclusive property of allelopathic agents in soil extracts but a function of all such factors acting independently and/or in a synergistic, additive, or antagonistic manner. Thus before allelopathic effects of extracts can be identified, the action of other physical and chemical properties of extracts must also be determined. Potential active properties of extract solutions can be determined by using appropriate controls and/or regression/response surface analysis to identify potential relationships between biological responses and soil extract characteristics.

Crimson clover (*Trifolium incarnatum* L. "Tibbee") and ivy-leaved morningglory (*Ipomoea hederacea* L.) were chosen as bioassay species since both species have been shown to be suppressed by wheat residues, soil extracts, and phenolic acids (Blum unpublished data; Liebl and Worsham 1983; Blum et al. 1992, 1993; Pue et al. 1995; Lehman and Blum 1997). Crimson clover is frequently used as a cover or forage crop and as a source of organic nitrogen while morningglory is a noxious weed. Water-autoclaved extracts of Cecil A soil samples (0–2.5 cm) from wheat stubble tilled under (*Triticum aestivum* L. "Coker 916")/soybean and fallow/soybean plots (conventional-till) did not inhibit germination or radicle/hypocotyl lengths of clover or morningglory. Extracts of Cecil A soil samples (0–2.5 cm) from wheat stubble/soybean plots (no-till) also did not inhibit germination, but inhibited radicle/hypocotyl lengths of both species. The importance of total phenolic acid (based on Folin & Ciocalteu's phenol reagent) and solute potential of the extracts on radicle/hypocotyl lengths of both species varied depending on which variable was placed first in the response surface analysis. This suggested that a co-linearity existed between total phenolic acid and solute potential of the extracts.

To characterize and isolate the effects of phenolic acids and solute potential, simulated wheat stubble/soybean soil extracts were created so that these variables could be manipulated independently. A range of concentrations of a mixture of 7 phenolic acids (0–393 μg/ml solution composed of 15% p-hydroxybenzoic, 10% caffeic, 17% vanillic, 10% syringic, 35% p-coumaric, 9% ferulic, and 4% sinapic; pH 5) mimicking wheat stubble/soybean soil extracts with a range of solute potentials (0–90 milliosmole) produced by different amounts of polyethylene glycol (PEG) or different strengths of Hoagland's solution were used in this modeling effort. These simulated soil extracts did not inhibit germination of crimson clover and morningglory or radicle/hypocotyl lengths of morningglory (Blum et al. 1992). The simulated soil extracts did reduce radicle/hypocotyl lengths of clover. Effects of the phenolic acids in the simulated extract varied with the osmoticum used. When the solute potential was increased by addition of PEG, the effects of the phenolic acid mixture declined (Fig. 3.10). When Hoagland's solution was used as the osmoticum, the actions of the phenolic acid mixture and solute potential were independent (Fig. 3.10). The difference in behavior of PEG and Hoagland's solution as an osmoticum indicated the presence of either interactions between PEG and phenolic acids or that the nature of the solute potential generated by PEG was much different than that of the Hoagland's solution. We suspect that the solute potentials generated by the Hoagland's solution were more consistent with field soils than

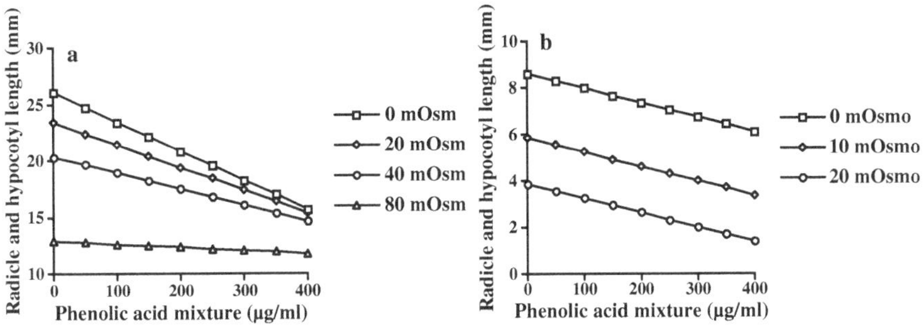

Fig. 3.10 Effects of a 7-phenolic acid solution modeled after phenolic acids found in wheat stubble/soybean (no-till) soil extracts (pH 5) on radicle and hypocotyl lengths of crimson clover as modified by solute potential of PEG (polyethylene glycol; **a**; $r^2 = 0.61$) and Hoagland's solution (**b**; $r^2 = 0.37$) based on freezing point depression (mOsm, milliosmoles) of solutions. The 7-phenolic acid mixture was composed of 10% caffeic acid, 9% ferulic acid, 35% *p*-coumaric acid, 15% *p*-hydroxybenzoic acid, 4% sinapic acid, 10% syringic acid, and 17% vanillic acid. Figures based on regressions from Blum et al. (1992). Plenum Publishing Corporation, regressions used with permission of Springer Science and Business Media

those generated by PEG (see below). What role, if any, individual nutrients in the Hoagland's solutions might have had was not determined. Previously it had also been demonstrated that the effects of phenolic acids were pH dependent (Harper and Balke 1981; Blum et al. 1985b; Shann and Blum 1987; Blum 1996; Lehman and Blum 1999) and thus, in retrospect, a range of pH values found in soils and soil extracts should have been included in this simulation.

Since physicochemical properties (e.g., pH, nutrients, organic particles/residues, and phenolic acid content) of soils are frequently uneven in distribution in field sites and at times co-linear, obtaining sets of extracts to be used in bioassays that are representative of a site and that have sufficient ranges for all the properties of interest is difficult even under the best of circumstances. The identification of important (active) soil properties in bioassays are further confounded by soil extracts which are borderline or not inhibitory (inactive). It seemed, therefore, that an estimate of biological activity for each extract would be much more useful in identifying the active properties of extract solutions since borderline and inactive extracts could be excluded. Serial dilutions of each soil extract, i.e., dose responses, were thus used to determine inhibition slopes for each extract. Significant linear slopes with a common intercept were found to be adequate to describe these slopes and were thus used as proxies for biological activity (Blum et al. 1992). Biological activity, i.e., slopes of extracts, were then regressed against initial undiluted extract properties (e.g., solute potential, pH, phenolic acid content, and ion content) to identify potentially active properties of soil extract solutions. In addition for adequately sampled sites, the frequencies of borderline, inactive, and active soil extracts could be used to determine the potential significance and frequency of plant–plant allelopathic interactions for those sites.

In a study utilizing crimson clover as bioassay species and slopes of soil extracts from randomly chosen wheat stubble/soybean soil samples, the relationships between mean radicle/hypocotyl lengths became more negative with increasing total phenolic acid and pH values (decreasing acidity) of extracts and the relationships became less negative initially before becoming more negative again with increasing solute potential of extracts; the more negative the slopes the more inhibitory the factors (Fig. 3.11). The effects of total phenolic acid, solute potential, and pH were independent. Other soil extract characteristics, such as NO_3^-, NH_4^+, K, and P, did not make a significant contribution to the observed changes in slopes. For all the soil extract bioassays (dose responses) run for this experiment, 97% of the extracts had negative significant slopes.

Additional comments regarding pH and solute potential effects are clearly warranted. Since decreasing acidities of phenolic acid solutions (i.e., increasing pH) have been shown to reduce the inhibitory activity of individual and mixtures of phenolic acids (Blum et al. 1985a, 1989), the independent and apparently additive effects of increasing pH (i.e., decreasing acidity) and total phenolic acid content of soil extracts on radicle/hypocotyl lengths were unexpected (Fig. 3.11). However, increasing recovery of phenolic acids from soils can be achieved when the pH of an extractant is increased (Whitehead et al. 1981; Dalton et al. 1983). In this instance the pH values of the extractants for the water-autoclave procedure were determined by the soil since water has no buffering capacity and the water-autoclave procedure has little impact on the extract pH. This suggested that the recovery of phenolic acids for soil samples with higher pH values were greater than soil samples with lower pH values and that the pH values of the soil extracts were co-linear with or a proxy for the concentrations of total phenolic acid recovered from the soil. However, since the inhibition of clover radicle/hypocotyl lengths by increasing pH and total phenolic acid were independent of each other and the range of the pH for the soil samples was ≤ 1 pH unit, other factors or inhibitors must have been involved. This suggests that total phenolic acid of the extract was really a proxy for a promoter/modifier/inhibitor

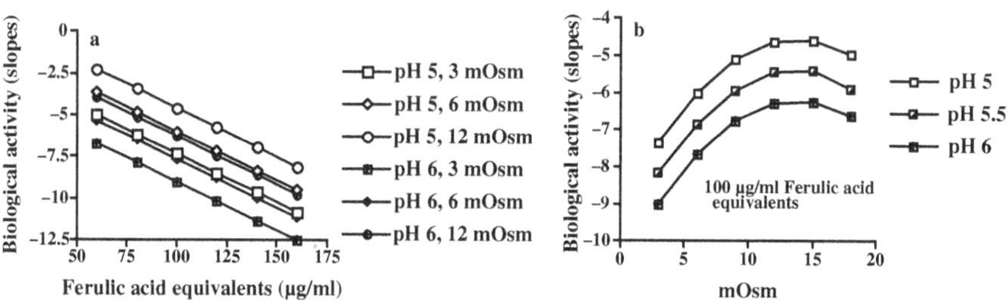

Fig. 3.11 Biological activity (slopes for radicle and hypocotyl lengths of crimson clover; $r^2 = 0.70$) from dose response studies (extract dilutions) of individual wheat stubble/soybean (no-till) soil extracts versus total phenolic acid (ferulic acid equivalence), pH, and freezing point depression (mOsm, milliosmoles) of original undiluted soil extracts. The more negative the biological activity the more inhibitory the factor. Figures based on regression from Blum et al. (1992). Plenum Publishing Corporation, regression used with permission of Springer Science and Business Media

complex dominated by inhibitors (see Section 2.6.1). The effects of solute potential appeared to be a possible example of hormesis (Fig. 3.11; Calabrese and Baldwin 2002). Hormesis is characterized by a biphasic dose response due to compensatory biological processes.

In summary extractions by the water-autoclave procedures in conjunction with the Folin & Cicoltaeu's phenol reagent method can provide relative or comparative estimates of phenolic acids present in wheat stubble/soybean, wheat stubble tilled under/soybean, and fallow/soybean soil samples. However, the Folin & Cicoltaeu's phenol reagent is reduced by a variety of other substances besides phenolic acids and, thus, in absolute terms the Folin & Cicoltaeu's phenol reagent method is likely to overestimate the actual phenolic acid concentrations in soil extracts (McAllister 1969; Box 1983). Furthermore, it is very likely that some of the reducing compounds and other compounds in the extracts, besides phenolic acids, were stimulatory, neutral, and/or inhibitory to clover radicle/hypocotyl growth and thus the total inhibition observed was very likely a result of a promoter/inhibitor complex and not purely the result of phenolic acids. However, since it has been demonstrated that neutral organic compounds, physiochemical factors, etc. can also increase or decrease the inhibition observed the effects were more likely due to a promoter/modifier/inhibitor complex (see Section 2.6.1; Blum et al. 1993; Pue et al. 1995). The utilization of appropriate species and slope analysis for individual soil extracts (dose response) appear to provide a useful measure of biological activity. The resulting slopes of a series of extracts for a treatment may be regressed against appropriate characteristics of each extract such as total phenolic acid content, solute potential, pH, etc. to study their role in determining the biological activity (slopes) of each extract. Thus, it appears that slopes, proxies for biological activity, of water-autoclaved soil extracts may be used to identify potential allelopathic activity and to determine how allelopathic activity may be modified by the actions and interactions of other environmental variables in soil extracts, a proxy for soil solutions.

3.4.3 Characterize How Cover Crop Residues in No-till Systems Affect Early Emergence of Broadleaf Weeds and to Establish and Characterize Potential Relationships Between Early Broadleaf Weed Seedling Emergence and the Physical and Chemical Environments Resulting from the Presence of Cover Crop Residues (Blum et al. (1997); Henry A Wallace Institute for Alternative Agriculture Inc, Summarized with Permission of Cambridge University Press)

A variety of glyphosate-desiccated cover crops (crimson clover (*Trifolium incarnatum* L. "Tibbee"), subterranean clover (*T. subterranean* L. "Mount Barker"), rye (*Secale cereale* L. "Abruzzi") and wheat (*Triticum aestivum* L. "Cocker 983")) were utilized in an attempt to create a range of phenolic acid concentrations and physicochemical environments in Cecil A soil of field plots (Blum et al. 1997). For one experimental period living cover crop biomass was also incorporated into the soil. Utilizing different seeding densities of cover crops to create a range of phenolic acid concentrations and physicochemical environments in soil plots was not successful with the rates used because vegetative reproduction (e.g., tiller and stolon formation) readily compensated for the lower seeding densities. Cover crops were planted by conventional tillage practices.

Weed seeds (morningglory (*Ipomoea hederacea* L.), red-root pigweed (*Amaranthus retroflexus* L.), and prickly sida (*Sida spinosa* L.)) were individually broadcast onto untilled (no-till) surface of the soil in subplots 3–5 month prior to their germination to allow for environmental selection to act upon the seed populations. Time for environmental selection was shorter for the tilled plots than the no-till plots because the broadcast of seeds was delayed until after plots were tilled. Germination of the original seed lots under laboratory conditions was 50–70% for morningglory and pigweed and 22–48% for prickly sida. Mean maximum weed seedling emergence in the field subplots of morningglory, pigweed, and prickly sida ranged from < 1 to 13% (mean = 4%) in the no-till systems and 3–16% (mean = 11%) in the tilled systems. Thus a large fraction of the seeds broadcast (300–2,500/m^2 depending on species or year) onto the subplots were either inactive (i.e., quiescent or dormant) or lost because of predation and/or mortality. High seed predation rates have been observed in other no-till systems (Brust and House 1988; Reader 1991; Brust 1994).

Weed seedling emergence in glyphosate-desiccated crimson clover, subterranean clover, rye and wheat residue no-till plots varied with year (1992 vs. 1993), date of glyphosate desiccation (April 1992, April and May in 1993), cover crop species, and weed species (Blum et al. 1997) when compared to no-cover crop no-till reference plots. In addition, in 1993 glyphosate-desiccated cover crops delayed initial emergence of morningglory, pigweed, and prickly sida seedlings when compared to no-cover crop reference plots (see Fig. 3.12 for one example; for other examples see Blum et al. 1997).

For simplicity let us compare the effects of desiccated cover crops on mean weed seedling numbers for the entire experimental periods (Fig. 3.13):

a. In 1992 the mean morningglory seedlings for the entire experimental period were suppressed by 54–75% in the presence of all cover crop residues. In 1993, mean morningglory seedlings were not modified or stimulated up to 46% by the clover cover crops and inhibited up to 57% by rye and wheat cover crops.
b. In 1992 mean pigweed seedlings were stimulated by clover cover crops ranging from 20 to 76% and inhibited by up to 96% by rye and wheat. In 1993, with one exception, mean pigweed seedlings were inhibited by all cover crops ranging from 21 to 92%.
c. Mean prickly sida seedlings, with one exception in 1992, were inhibited by all cover crops for both years ranging from 15 to 81%, and

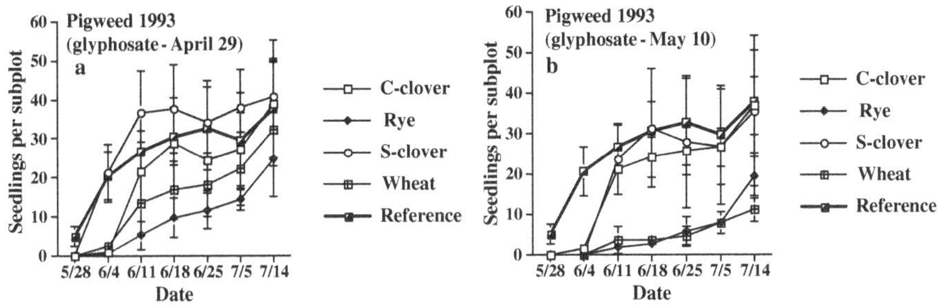

Fig. 3.12 The number of pigweed seedlings in cover crop and reference plots for the 1993 experimental period in no-till Cecil A soil. Glyphosate desiccation of cover crops occurred on April 29 (**a**) and May 10 (**b**). Where C equals crimson, S equals subterranean and reference equals no-cover crop plots. Figures reproduced from Blum et al. (1997). Henry A Wallace Institute for Alternative Agriculture Inc, figures used with permission of Cambridge University Press

d. Suppression of weed seedling emergence was greater when cover crops were desiccated in May, which was closer to the time of seedling emergence, than in April of 1993. However, it is not clear if that difference was entirely due to timing or due to difference in inhibitors of cover crops (i.e., cover crops were 1 month older).

Why the difference between 1992 and 1993? The following differences in weather and field management may have been directly or indirectly responsible for the differences in weed seedling emergence.

a. Mean April–August air temperature was 1.4 and 0.7°C above normal (22°C) in 1992 and 1993, respectively.
b. Rainfall for April–August was 7.8 and 22.6 cm below normal (57 cm) in 1992 and 1993, respectively. June of 1993 was so dry that plots were irrigated (2.5 cm) with an overhead sprinkler system to prevent the loss of the experiment.
c. Although locations of treatment plots each year were chosen at random there may have been differences due to cover crop "carry-over" from 1 year to the next since the experiments for the 2 years were carried out in the same field. For example the soil nitrate-N levels for clover no-till plots in 1993 (mean values from 22 to 50 μg/g soil) were considerably higher than for the 1992 plots (mean values from 5 to 9.5 μg/g soil; Blum et al. 1997).
d. Wheat and rye plots were fertilized in March 1993. Their nitrate-N levels ranged from 1 to 2 μg/g soil in 1992 and 11–22 μg/g soil in 1993, and
e. Cover crops were desiccated at different time in 1992 (middle April) and 1993 (late April and early May). For additional differences see Section 3.4.4.

Weed seedling emergence was stimulated when living cover crop biomass was incorporated into the soil. Stimulation for morningglory ranged from 172 to 409%

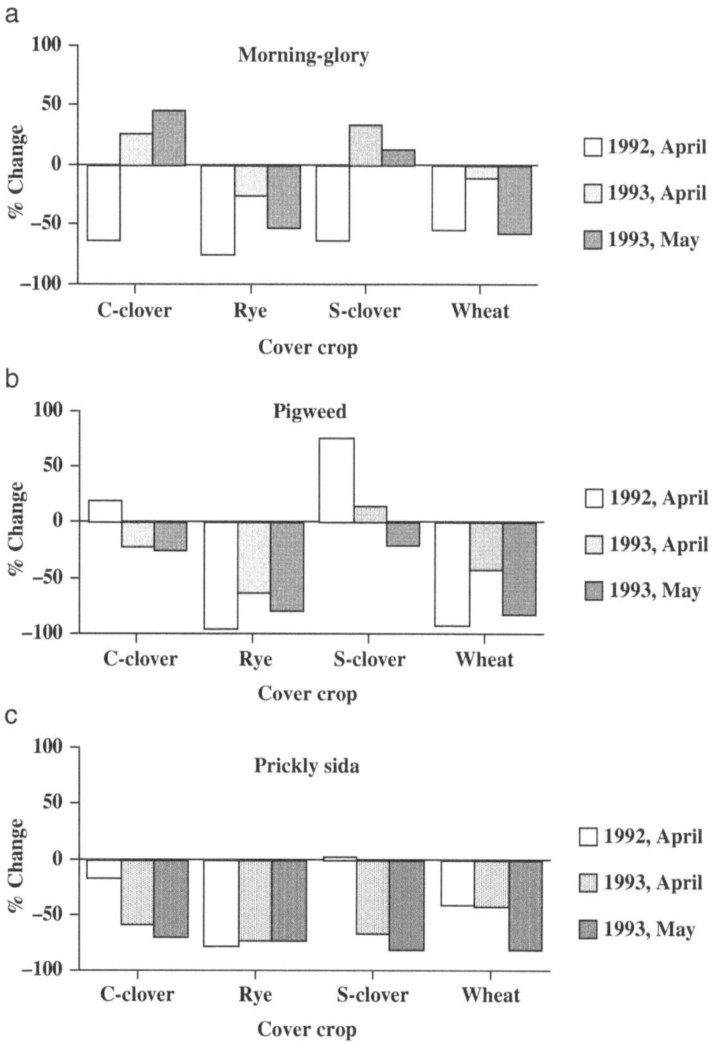

Fig. 3.13 Percent change in mean seedling numbers of morningglory (**a**), pigweed (**b**), and prickly sida (**c**) due to presence of desiccated cover crops for the 1992 and 1993 experimental periods in no-till Cecil A soil, where C equals crimson and S equals subterranean. Figures based on data from Blum et al. (1997). Henry A Wallace Institute for Alternative Agriculture Inc, data used with permission of Cambridge University Press

and prickly sida ranged from 34 to 62%. Response of pigweed seedlings to living biomass incorporation could not be determined since reference plots were over-whelmed by seedlings from a natural weed population close to the plot. The stimulation of morningglory and prickly sida was very likely due to an increase of "safe" germination sites, the destruction of the surface layers of the soil, site of highest concentrations of potential inhibitors (Putnam and DeFrank 1983), and/or by bringing weed seeds from the lower soil horizons to the surface. In addition the shorter time interval for natural selection of the weed populations (plots seeded only after cover crops were incorporated) and high soil nitrate-N levels (mean range 65–81 μg/g soil) may also have played a role.

Covariate, correlation, and principle component analyses did not identify any significant relationships between seedling emergence and soil physical and chemical characteristics determined (e.g., soil total phenolic acid, nitrate-N, pH, moisture, and temperature; for data see Section 3.4.4). This was not entirely surprising since the bulk-soil data collected were very limited. Clearly much more precise, extensive, and continuous physicochemical data will be required to establish any such relationships. However, even collecting more detailed bulk-soil physicochemical data may be of limited value, since the actions of environmental factors normally operate on the micro-site level, e.g., the rhizosphere/rhizoplane, and bulk-soil characteristics are not necessarily representative of what is occurring at the rhizosphere/rhizoplane level. As an aside, it may be noteworthy to add that Blum et al. (1997) observed that the cover crops provided only weak control of a "natural" infestation of broadleaf signal grass (*Brachiaria platyphylla* [Grisebach] Nash) suggesting that control of monocotyledonous species by legume and grain cover crops may be a problem. Weaker suppression of grasses compared to broadleaf weeds by cover crops has also been observed by other researchers (Shilling et al. 1986; Worsham 1990; Wickliffe 1999).

In summary clearly the presence and management of cover crops (e.g., type of cover crop, amount of cover crop biomass and residues, time of glyphosate desiccation, level and frequency of irrigation and precipitation, and presence or absence of tillage) can substantially influenced seedling emergence of morningglory, pigweed and prickly sida in either a positive or negative manner. Negative effects, such as a delay in initial seedling emergence were short lived, on the order of a week or two. Once initiated, weed seedling emergence continued at rates ranging from reduced to greater than that observed in no-till no-cover crop reference plots. Thus seedling emergence over the experimental periods, depending on weed species and management procedures, can potentially range from suppression to stimulation when compared to the no-till no-cover crop reference plots. It is also clearly evident from this study that morningglory, pigweed and prickly sida respond very differently to the presence of cover crops and cover crop residues and that their responses can vary considerably from year to year. However, the biotic and physicochemical factors responsible for these positive or negative responses of individual weed species were not clearly evident for this field study.

To provide a proper context the following needs to be stated in regard to weed seedling emergence observed in these two field studies. No crop plants (e.g., corn or soybean) were included in these studies and the effects of crops (e.g., shading of crop canopies and competition for moisture) on weed seedling emergence over the two experimental periods were thus absent. The differences in weed seedling numbers for the reference and cover crop treatments over these two experimental periods were thus largely a product of the absence/presence of cover crop tissues/residues. The inclusion of crops would have made these studies more meaningful in terms of weed management for crop production, but such inclusion of crops would have substantially increased the number of treatments (i.e., plus and minus crops) required to separate the effects of cover crops and crops on weed emergence. Recall that the primary goal was to determine the effects of desiccated cover crop residues on weed seedling emergence and not weed management.

3.4.4 Characterize Cover Crops and Cover Crop Residues and How These May Potentially Modify the Soil Environment (Blum et al. (1997); Henry A Wallace Institute for Alternative Agriculture Inc, Summarized with Permission of Cambridge University Press)

When determining the potential influence of cover crops on the physicochemical state of Cecil A no-till soils or tilled soils, there are a variety of factors that are of interest. Among them are: above- and below-ground cover crop biomass, morphology of cover crops (as related to shading and root distribution), soil temperature, soil moisture, soil compaction, soil pH, chemical composition of cover crops before and after glyphosate desiccation (e.g., C/N ratio and phenolic acid content), soil C/N ratio, soil nitrate-N, and soil phenolic acid content.

3.4.4.1 Cover Crop Biomass

In addition to the weather (see Section 3.4.3) cover crop management varied in 1992 and 1993. In 1992 glyphosate was applied to all plots on April 16 and in 1993 glyphosate was applied to subsets of plots on April 29 and May 10 (Blum et al. 1997). Also in 1993 living biomass was incorporated into the soil for another subset of plots on April 29 and wheat and rye plots were fertilized with nitrogen on March 31. As might be suspected biomass of cover crops and their residues after glyphosate desiccation were considerably different not only between 1992 and 1993 but also among the different cover crops, glyphosate-desiccation times, and the no-till and living biomass tillage treatments in 1993. For the 1992 experimental period (biomass was collected on May 12 and June 19; April 16 glyphosate desiccation) mean dry weights of surface cover crop residue were 863 $\pm$ 83 g/m^2 for crimson clover, 1,114 $\pm$ 100 g/m^2 for rye, 461 $\pm$ 34 g/m^2 for subterranean clover, and 954 $\pm$ 83 g/m^2 for wheat (Blum et al. 1997). On average the cover crop residues declined by 30% from May 12 to June 19. For the 1993 experimental period (biomass was collected on May 28 and June 28) and an April 29 glyphosate desiccation, these values were 408 $\pm$ 81 g/m^2 for crimson clover, 435 $\pm$ 48 g/m^2 for rye, 422 $\pm$ 35 g/m^2 for subterranean clover, and 497 $\pm$ 52 g/m^2 for wheat (Blum et al. 1997). For the glyphosate desiccation of May 10 these values were 159, 96, 101, and 6 g/m^2 higher than the April 29 desiccation, respectively. On average the cover crop residues for all surface residues declined by 16% from May 28 to June 28. Mean dry weight of living biomass on April 29, just before living biomass was incorporated into the soil, was 844 ± 30 g/m^2 for crimson clover, 563 ± 41 g/m^2 for rye, 800 ± 40 g/m^2 for subterranean clover, and 407 ± 45 g/m^2 for wheat (Blum et al. 1997). Although belowground root dry weight was not determined for either year, we suspect that ≥ 50% of aboveground dry weight biomass would be a reasonable estimate since root/shoot ratios of 0.5–1 in grasses and legumes are not uncommon (Davidson 1969a, b; Brouwer 1983).

3.4.4.2 Cover Crop Morphology

Crimson and subterranean clovers are composed primarily of leaves, petioles, and runners (stolons) while wheat and rye are composed primarily of leaves and stems. The clovers have a taproot and primary, secondary, and tertiary roots while the grasses have primarily adventitious roots. Upon desiccation, crimson clover aboveground residue either remained erect or formed dense tangles of petioles and runners resulting in a mosaic of open and covered soil surfaces. Subterranean clover residue formed a very thin dark brown mat that covered the surface. Rye residue quickly formed a thick dense mat that covered most of the soil surface. Most of the wheat residue remained initially erect resulting in considerable open soil surface area. The amount of open soil surface area, however, varied with wheat density and time of glyphosate desiccation. As wheat stalks fell over and/or shattered after glyphosate desiccation, the open soil surface area was reduced with time and eventually eliminated.

3.4.4.3 Solar Radiation, Temperature, Moisture, Compaction, and pH

The aboveground characteristics of the cover crops influenced solar penetration and thus mean soil temperature and moisture. For a quick survey at noon on June 8, 1996 we found a > 85% reduction in blue (400–600 nm) and red (600–700 nm) solar radiation at the soil surface underneath approximately 930 g/m^2 of standing or cut wheat shoot residues (Blum et al. 2002). The solar radiation at the soil surface of the reference plots on that day was 190 ± 6 μW/cm^2/nm.

Mean soil temperature readings at 2.5 cm on two clear sunny afternoons in early and late June of 1992 for the no-till reference plots (no-cover crop plots) were 38 ± 1°C and ranged from 28 ± 0.4°C for rye plots to 34 ± 0.4°C for subterranean clover plots (Blum et al. 1997). Soil temperature readings at 2.5 cm for 4 days during May 28–June 25 of 1993 were similar except that the mean soil temperatures were higher. Reference plots were 41 ± 1.0°C and cover crop plots for April glyphosate desiccation ranged from 33 ± 0.8°C for rye to 36 ± 1.1°C for wheat. For the May glyphosate desiccation these values ranged from to 32 ± 1.2°C for subterranean clover to 35 ± 1.0°C for wheat. The mean soil temperatures for the incorporated cover crop plots were slightly lower (0.7°C) than the tilled reference plots (tilled reference plots were 40 ± 1.7°C and ranged from 39 ± 1.0°C to 41 ± 1.1°C for incorporated cover crop plots; Blum et al. 1997).

Mean soil moisture values in 1992, based on 4 sampling times, were 0.109 ± 0.005 g water/g soil for the no-till reference plots and ranged from 0.136 ± 0.003 g water/g soil for subterranean clover plots to 0.163 ± 0.003 g water/g soil for rye plots (Blum et al. 1997). In 1993, mean soil moisture values were 0.093 ± 0.008 g water/g soil for the no-till reference plots. For the April glyphosate desiccation the values ranged from 0.117 ± 0.008 g water/g soil for wheat plots to 0.127 ± 0.010 g water/g soil for subterranean clover plots. For the May glyphosate desiccation the values ranged from 0.113 ± 0.010 g water/g soil for wheat plots to 0.126 ± 0.008 g water/g soil for crimson and subterranean clover plots. The mean soil moisture values were 0.08 ± 0.017 g water/g soil for the tilled reference plots and ranged from 0.087 ± 0.013 g water/g soil for wheat plots to 0.097 ± 0.013 g water/g soil for subterranean clover plots (Blum et al. 1997).

Soil compaction determined for no-till plots on July 5 in 1993 with a soil penetrometer found that only reference plots (28 ± 2 mm penetration) and rye plots (40 ± 2 mm penetration) were significantly different (Blum et al. 1997). Dramatic differences would not be expected since these were all one season no-till plots. No significant differences were observed for the incorporated cover crop plots. Minimum and maximum penetration for these plots ranged from 40 ± 3 (wheat) to 53 ± 4 (crimson clover) mm. Soil pH, determined approximately biweekly for the experimental period, was not modified by the presence of the various cover crops. Mean pH values ranged from 4.8 to 5.2.

3.4.4.4 C/N Ratios and Nitrate-N

Lehman and Blum (1997) found that the mean C/N ratios of cover crop residues were 51, 32, 15, and 12 for rye, wheat, crimson clover and subterranean clover, respectively. The C/N ratios of the soils, however, were not significantly modified in 1993 (only year determined). The values ranged from 9 to 19 with a mean of 13.5 ± 0.13 (Blum et al. 1997). Both Iritani and Arnold (1960) and Harmsen and Van Schreven (1955) have suggested that a C/N ratio of 20 is the approximate threshold between mineralization and immobilization of N in soil by microorganisms. The mean value of 13.5 for these soils suggests that in most instances the soils in 1993 were probably not nitrogen limited.

Soil nitrate-N values, however, were modified by the presence of the cover crops. In 1992 the means for soil nitrate-N determined at two points in time (May 12 and June 19) were as follows: no-till plots without cover crops (reference plots) 3.25 ± 1.11 μg/g soil, crimson clover plots 4.82 ± 0.85 μg/g soil, subterranean clover plots 9.44 ± 1.58 μg/g soil, rye plots 1.71 ± 0.04 μg/g soil, and wheat plots 0.80 ± 0.16 μg/g soil (Blum et al. 1997). In 1993 (April desiccation) the means for soil nitrate-N determined at three points in time (May 31, June 14, and July 5) were as follows: no-till plots without cover crops (reference plots) 18 ± 5 μg/g soil, crimson clover plots 34 ± 7 μg/g soil, subterranean clover plots 50 ± 12 μg/g soil, rye plots 13 ± 3 μg/g soil, and wheat plots 22 ± 5 μg/g soil (Blum et al. 1997). For May desiccation these values were as follows: no-till plots without cover crops (reference plots) 18 ± 5 μg/g soil, crimson clover plots 22 ± 5 μg/g soil, subterranean clover plots 50 ± 13 μg/g soil, rye plots 11 ± 3 μg/g soil, and wheat plots 14 ± 5 μg/g soil. The differences in nitrate-N between 1992 and 1993 were due partly to "carry over" from the 1992 season to the 1993 season and fertilization of rye and wheat plots in March of 1993. Mean soil nitrate-N for the incorporated living biomass plots were as follows: tilled plots without cover crops (reference plots) 67 $\pm$ 2 μg/g soil, crimson clover plots 79 $\pm$ 3 μg/g soil, subterranean clover plots 95 $\pm$ 4 μg/g soil, rye plots 81 $\pm$ 3 μg/g soil, and wheat plots 65 $\pm$ 1 μg/g soil (Blum et al. 1997).

3.4.4.5 Phenolic Acids in Cover Crops

One source of phenolic acids in no-till cover crop agricultural systems is the plant residues on and in the soil. As would be expected phenolic acid content of plant residues will vary with cover crop species, stage of development, type of tissue, time of desiccation, stage of decay, rainfall events (timing, frequency, and duration), extraction procedure used, etc. Thus phenolic acid content of plant residues is dynamic and to a large extent difficult to predict. The order of total phenolic acid content of freeze-dried surface cover crops, before glyphosate desiccation, based on water-autoclave extractions were as follows: subterranean clover, rye, crimson clover, and wheat (Lehman 1993). Concentrations in ferulic acid equivalence ranged from 14 $\pm$ 0.22 mg ferulic acid equivalents/g wheat residue to 33 $\pm$ 0.19 mg/g subterranean clover residue. Concentrations for rye and crimson clover were 17 $\pm$ 0.35 and 15 $\pm$ 1.06 mg/g residue, respectively. Total phenolic acid content of these cover crop extracts was estimated using the Folin & Cicoltaeu's phenol reagent procedure. Estimates using the PVP method which also utilizes the Folin & Cicoltaeu's phenol reagent (Anderson and Todd 1968; Lehman and Blum 1997) were linearly related to the estimates obtained by the water-autoclave procedure. On average the values for the water-autoclave procedure were 1.6 times greater than the values for the PVP method. These relationships will of course vary with organic and inorganic composition of extracts. The PVP method is generally considered to be more reliable because the PVP method reduces, if not eliminates, the role of other reactive organic compounds (McAllister 1969; Anderson and Todd 1968; Box 1983). However, since the unit of ferulic acid equivalents was based on the color development of one phenolic acid, i.e. ferulic acid, and intensity of color development with the Folin & Ciocalteu's phenol reagent varies with phenolic acid, such values can only be used for comparative purposes (Blum et al. 1991). Please also note that since organic and inorganic composition of small grains and clovers vary dramatically direct comparisons of total phenolic acid content between small grains and clovers may not be appropriate because the Folin & Ciocalteu's phenol reagent is reduced not only by phenolic acids but also by some other organic and inorganic compounds including some amino acids that are higher in clovers (McAllister 1969; Box 1983).

Total phenolic acid content of the cover crops taken at monthly intervals from the field after glyphosate desiccation declined over time. However, the decline of total phenolic acid content for rye and wheat residues was not evident until 2 months after glyphosate desiccation (Lehman 1993). After 4 months the total phenolic acid content of the cover crops had declined by 31, 36, 38, and 56% for wheat, crimson clover, rye, and subterranean clover, respectively (Lehman 1993). Estimates of ferulic acid for sterilized (autoclaved) cover crop residues extracted with 0.25 M citrate and quantified by HPLC analysis were 25 μg/g for wheat, 100 μg/g for crimson clover, 40 μg/g for rye, and 20 μg/g for subterranean clover (Blum 1997). For non-sterilized residues the values were 200, 210, 85, and 25, respectively. The difference between the sterilized and non-sterilized residues suggests that dry (no water added) autoclaving of residues leads to irreversible sorption and/or polymerization of ferulic acid based on citrate extractions.

Wheat straw ("Coker 916") from half buried litter bags that was analyzed more extensively for individual phenolic acids contained the following concentrations in μg/g straw ± standard error based on water-autoclave extracts and HPLC analysis: *p*-coumaric acid 261 ± 26, sinapic acid 97 ± 23, caffeic acid 62 ± 4, ferulic acid 44 ± 3, syringic acid 40 ± 7, vanillic acid 33 ± 3, and *p*-hydroxybenzoic acid 13 ± 1 (Fig. 3.8; Blum et al. 1991). The sum of the 7 phenolic acids was 549 ± 32 μg/g residue. The total concentrations for each individual phenolic acid are actually much higher since water-autoclave extractions, or for that matter water, neutral EDTA, or citrate extractions, recover primarily only the soluble and/or reversibly sorbed phenolic acids. Thelander (1985) using much more rigorous extraction procedures determined that wheat, barley and oat straw contained 3 mg/g of *p*-coumaryl and 2 mg/g ferulyl most bound to hemicellulose and cellulose in cell walls. That is roughly 11 and 45 times greater, respectively, than what was extracted by the water-autoclave extraction. Phenolic acid content of roots was not determined. Wu et al. (2000b) analyzing 17-day old wheat seedlings observed significant correlation between phenolic acid content (e.g., cinnamic and benzoic acids) of shoots and roots but that the phenolic acid content of roots was higher than in shoots in many instances. For example for the wheat cultivar "Tasman" the root/shoot ratios were 4.6 for *cis*-ferulic acid, 1 for *trans*-ferulic acid, 2.8 for *cis*-*p*-coumaric acid, 4.0 for *trans*-*p*-coumaric acid, 2.6 for *p*-hydroxybenzoic acid, 1.8 for vanillic acid, and 0.21 for syringic acids. However, freeze-dried root tissues of mature Southern States 555 wheat plants ground and mixed into soil were found to be less inhibitory to pigweed seedling emergence than freeze-dried shoot tissues (Staman et al. 2001). Five mg/g soil of shoot and root tissue inhibited pigweed seedling emergence by 66 and 10%, respectively. The root/shoot ratio for total phenolic acid content was 0.76.

3.4.4.6 Phenolic Acid in Soils

All total phenolic acid values in ferulic acid equivalence were determined by water-autoclave extraction and the Folin & Cicoltaeu's phenol reagent approximately biweekly during the 1992 and 1993 experimental periods. Mean total phenolic acid for the upper 2.5 cm of the soil in 1992 were as follows: plots without cover crops (reference plots) 41 ± 1 μg/g soil (mean ± standard error), crimson clover plots 55 ± 1 μg/g soil, subterranean clover plots 58 ± 1 μg/g soil, rye plots 50 ± 1 μg/g soil, and wheat plots 45 ± 1 μg/g soil (Fig. 3.14; Blum et al. 1997). In 1993 soil total phenolic acid values for the upper 2.5 cm of the soil were higher than in 1992 (Fig. 3.14; Blum et al. 1997). The values for the April glyphosate desiccated plots were: plots without cover crops (reference plots) 62 ± 2 μg/g soil, crimson clover plots 83 ± 4 μg/g soil, subterranean clover plots 83 ± 3 μg/g soil,

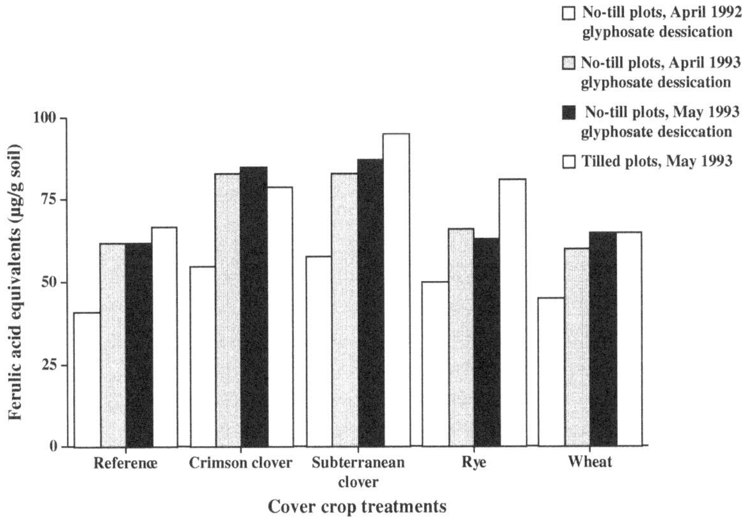

Fig. 3.14 Mean total phenolic acid (ferulic acid equivalents) content of 0–2.5 cm Cecil soil samples taken during the 1992 and 1993 growing season for reference plots (no-cover crop) and cover crop plots. In 1992 cover crops were desiccated with glyphosate in April. In 1993 cover crops were desiccated with glyphosate at two time periods (April and May) and living biomass was tilled into plots in May. The absence of standard error bars indicates that the error bars are too *small* to be visible. Figure based on data from Blum et al. (1997). Henry A Wallace Institute for Alternative Agriculture Inc, data used with permission of Cambridge University Press

rye plots 66 ± 2 µg/g soil, and wheat plots 60 ± 2 µg/g soil. The values for the May glyphosate-desiccated plots were slightly higher (i.e., 2–5 µg/g soil) than for the April glyphosate plots. The total phenolic acid values for the incorporated living biomass plots in 1993 were: plots without cover crops (reference plots) 67 ± 2 µg/g soil, crimson clover plots 79 ± 3 µg/g soil, subterranean clover plots 95 ± 4 µg/g soil, rye plots 81 ± 3 µg/g soil, and wheat plots 65 ± 1 µg/g soil. Please note that since organic and inorganic composition of small grains and clovers vary dramatically direct comparisons of total phenolic acid content between small grains and clover soils may not be appropriate because the Folin & Ciocalteu's phenol reagent is reduced not only by phenolic acids but also by some other organic and inorganic compounds including some amino acids that are higher in clovers (McAllister 1969; Box 1983).

3.4.4.7 Relationships to Weed Seedling Emergence

At this point some comments regarding the modification of the soil physical and chemical environments by cover crops and weed seedling emergence appear appropriate. In spite of the fact that covariate, correlation and principle component analyses did not identify any significant relationships between seedling emergence and bulk soil physical and chemical characteristics (e.g., soil total phenolic acid, nitrate-N, pH, moisture, and temperature), the observed stimulation and/or inhibition of weed seedling emergence in the presence of glyphosate-desiccated small grain and clover cover crops were certainly influenced indirectly, if not directly, by the observed changes of the physicochemical and/or the biotic environments of the bulk soil (Blum et al. 1997, 2002). However, stimulation or inhibition of weed seedling emergence is not regulated by the bulk soil but by the immediate environment surrounding weed seeds and/or the rhizosphere/rhizoplane of weed seedling roots (Rovira 1991; Walker et al. 2003; Sylvia et al. 2004; Blum and Gerig 2005; also see Section 2.4.4). Since seeds, seedling roots, and associated microbes can modify their immediate surroundings, they may be partially or largely uncoupled from the environment of the bulk soil. Bulk-soil data, however, do provide some insight as to the types of changes that may have occurred within the weed seed zone and the rhizosphere/rhizoplane of weed seedling roots and thus provide a basis for designing experiments to determine potential cause and effect. Of particular interest were the potential roles and interactions of soil phenolic acids, nitrate-N, temperature, and moisture, the action of phenolic acid-utilizing soil microbes, and the individual roles of shoot and root residues (see Sections 3.4.5, 3.4.6 and 3.4.7).

3.4.4.8 Summary

The presence of living and desiccated cover crops modify the physicochemical and biotic environments of the soil and function as a reservoir of allelopathic agents, such as phenolic acids, and other organic and inorganic compounds. Their impact varies with biomass, morphology, and chemical composition which all change over time as living cover crops grow and desiccated cover crops are leached by rain events, fragmented and decomposed. The presence of glyphosate-desiccated cover crops modified the solar radiation at the soil surface, soil temperature, moisture, nitrate-N and phenolic acid content of the soil. In general, solar radiation in the red and blue range was reduced (> 85%), soil moisture (0–2.5 cm) was increased (mean range of plus 0.02–0.05 g water/g soil), and soil temperature (2.5 cm) was reduced (mean range of minus 4–10°C) when cover crops were present. Soil nitrate-N (0–2.5 cm) was higher for the clover cover crop plots (mean range of plus 1.6–6.2 μg/g soil in 1992 and plus 4–32 μg/g soil in 1993) when compared to the no-till reference plots. The highest concentrations were observed for the subterranean clover plots. Soil nitrate-N of the small grain cover crop plots were generally reduced (mean range of minus 1.5–7 μg/g soil; one exception was plus 4 μg/g soil) when compared to no-till reference plots. These findings are consistent with what others have observed (Kimber 1973; Fester and Peterson 1979; Worsham 1989). The presence of incorporated living biomass in 1993 did not modify solar radiation at the soil surface, marginally modified soil temperature and moisture, and increased soil nitrate-N (mean range of plus 12–28 μg/g soil) for all but wheat plots when compared to the tilled reference plots. Total phenolic acid content in ferulic acid equivalence was generally modified by the presence of both glyphosate-desiccated and incorporated living cover crops. In 1992 for desiccated cover crops the soil total phenolic acid content was equal to or higher for the small grain cover crop (mean range of plus 4–9 μg/g soil) and clover cover crops (mean range of plus 21–25 μg/g soil) when compared to the no-till reference plots. In 1993 for desiccated cover crops the soil total phenolic acid content was equal to or higher for the small grain cover crops (mean range of minus 2 to plus 4 μg/g soil) and for clover cover crops (mean range of plus 21–25 μg/g soil) when compared to no-till reference plots. For incorporated cover crops in 1993 the soil total phenolic acid content was equal to or higher for the small grain cover crops (mean range of minus 3 to plus 14 μg/g soil) and for clover cover crops (mean range of plus 12–28 μg/g soil) when compared to tilled reference plots. Soil total phenolic acid content of all plots was 21–28 μg/g soil higher in 1993 than in 1992. The presence of cover crops did not or only marginally modify pH, compaction, and C/N ratios of soils. No significant relationships between weed seedling emergence and changes in the physicochemical environment due to cover crops were identified.

3.4.5 Determine Under Controlled Conditions How Effects of Shoot Cover Crop Residues Taken from the Field Change with Time After Desiccation and How Such Effects Are Modified By Temperature, Moisture, and Nitrogen Levels (Lehman and Blum (1997); Summarized with Permission of International Allelopathy Foundation)

A number of studies have shown that inhibitory activity of residues in the field decline over time (Glass 1976; Bhowmik and Doll 1983; Einhellig 1987; Blum et al. 1992; Yenish et al. 1995; Bonanomi et al. 2006). We found the following:

a. All shoot tissues of rye, wheat ("Coker 983"), crimson clover, and subterranean clover collected prior to glyphosate desiccation and lyophilized, ground, and mixed into Cecil A soil-sand mixture inhibited the emergence of ivy-leaf morningglory and pigweed seedlings under laboratory conditions (Lehman and Blum 1997). For example 10 mg tissue/g soil of crimson clover, wheat, subterranean clover, and rye inhibited morningglory seedling emergence by 28, 17, 33, and 22%, respectively. Pigweed seedling emergence was inhibited by 18, 24, 30, and 30%, respectively. Total phenolic acid content of cover crop tissues was as follows in mg ferulic acid equivalence/g tissue: crimson clover 15.37 ± 1.06, wheat 13.81 ± 0.22, subterranean clover 33.12 ± 0.19, and rye 16.82 ± 0.35 (Lehman 1993).
b. For residues collected 1 month after glyphosate desiccation only shoot crimson clover and wheat residues were inhibitory. Ten mg/g soil of crimson clover and wheat residues inhibited morningglory seedling emergence by 15 and 30%, respectively (Lehman and Blum 1997). Pigweed seedling emergence was inhibited by 16 and 23%, respectively. Total phenolic acid content of cover crop residues was as follows in mg ferulic acid equivalence/g residue: crimson clover

8.45 ± 0.38 and wheat 18.88 ± 0.17 (Lehman 1993). The values for rye and subterranean clover were 17.36 ± 0.15 and 17.18 ± 0.46, respectively, and

c. For residues collected more than 1 month after glyphosate desiccation only wheat residues inhibited pigweed seedling emergence. Ten mg/g soil of wheat residues collected 2, 3, and 4 months after the glyphosate desiccation inhibited pigweed seedling emergence by 19, 15, and 19%, respectively (Lehman and Blum 1997). Total phenolic acid content of wheat cover crop residues in mg ferulic acid equivalence/g residue was 12.83 ± 0.08, 11.89 ± 0.06, and 9.58 ± 0.01, respectively (Lehman 1993). Total phenolic acid contents for clovers and rye during this same time interval were the following: crimsons clover – 10.46 ± 0.04, 8.58 ± 0.26, and 9.87 ± 0.04; subterranean clover – 14.16 ± 0.10, 15.61 ± 0.06 and 14.5 ± 0.04; and rye – 13.80 ± 0.02, 11.74 ± 0.07, and 10.36 ± 0.07.

Please note that since organic and inorganic composition of small grains and clovers vary dramatically direct comparisons of total phenolic acid content between small grains and clovers may not be appropriate because the Folin & Ciocalteu's phenol reagent is reduced not only by phenolic acids but also by some other organic and inorganic compounds including some amino acids that are higher in clovers (McAllister 1969; Box 1983). A quick comparison of the data for inhibition of morningglory and pigweed seedling emergence and the data for total phenolic acid content of cover crop tissue/residues suggested that the relationships between inhibition and total phenolic acid for the two weed species were not very robust. In fact total phenolic acid content of tissues/residues was only significantly related to the inhibition slopes (see Section 3.4.2) of morningglory seedling emergence ($r^2 = 0.48$) and not the inhibition slopes of pigweed seedling emergence (Lehman and Blum 1997). Clearly other things must have been involved in the inhibition of seedling emergence besides phenolic acids, e.g., environmental factors and/or other inhibitors. It has been demonstrated that plant–plant allelopathic effects of individual and mixtures of phenolic acids may be altered by the presence of other organic compounds, by physical factors of the environment (e.g., pH, temperature, soil moisture, solute potential, and nitrogen levels in soil) and by microbial activity (Turner and Rice 1975; Glass 1976; Stowe and Osborn 1980; Sparling et al. 1981; Sparling and Vaughan 1981; Bhowmik and Doll 1983; Hall et al. 1983; Einhellig 1987; Blum and Shafer 1988; Blum et al. 1989; Shafer and Blum 1991; Blum et al. 1992, 1999a, b; Pue et al. 1995; Blum 1998, 2004; Lehman and Blum 1997; Staman et al. 2001). Thus the question, how are the effects of cover crop residues on weed seedling emergence modified or altered by such factors as pH, temperature, soil moisture, and/or nitrogen levels in the soil? The role of solute potential was addressed in Section 3.4.2 and the role of microbial activity will be addressed in Section 3.4.7.

Blum et al. (1997) found that soil pH values of no-till systems were not significantly modified by the presence of glyphosate-desiccated rye, wheat, crimson clover and subterranean clover cover crops. In this field study a substantial portion (estimated to be roughly 50%) of the cover crop residues was located on the soil surface and thus only had an indirect impact on soil pH by way of leachates generated by rainfall events. In laboratory studies, particularly when residues are ground and mixed with soil, that is not the case (Blum 1999). Lehman and Blum (1997) found that mixing ground rye, wheat, crimson clover and subterranean clover shoot tissues/residues into soil changed the pH of the soil. For example, 10 mg/g soil of rye, wheat, crimson clover and subterranean clover shoot tissues collected prior to glyphosate desiccation raised the soil pH (pH of control soil was 5.1) by 0.4, 0.2, 2.5, and 1.5 pH units, respectively. Residues collected 2 month after glyphosate desiccation raised the soil pH by 0.6, 0.3, 0.5, and 1.5 units, respectively. How important these soil pH changes may have been to the observed inhibition of morningglory and pigweed seedling emergence was not determined. We do know from nutrient and soil culture studies that the dose of phenolic acids required for a given inhibition increases as the pH of the medium increases (becomes less acidic; Blum et al. 1985b, 1989). We also know that soil pH, particularly between pH 4.5 and pH 6 can influence the available/active concentrations of phenolic acids in soil. For example Blum et al. (1992) observed that increasing soil extractant pH appeared to increase the recovery of phenolic acids by the water-autoclave procedure. This suggested that higher amounts of phenolic acids may have been available in soils with higher soil pH values, however on the other hand, more of the phenolic acids would also have been negatively charged, i.e., less active (Harper and Balke 1981; Blum et al. 1985b; Shann and Blum 1987; Lehman and Blum 1999).

In the absence of cover crop residues Lehman and Blum (1997) found that morningglory seedling emergence was not affected by temperature (25/21–35/31°C, day/night), ammonium nitrate (0–14 μg N/g soil), or soil moisture (0.093–0.173 g water/g soil) utilized in their study. Emergence of pigweed seedlings was linearly related to increasing temperature, strongly inhibited by higher soil moisture (> 0.16 g water/g soil), but not modified by nitrogen levels utilized (Fig. 3.15). However, the effects of cover crop residues on weed seedling emergence were at times modified by interactions of temperature, moisture, and/or nitrogen. Two examples:

a. The inhibition of morningglory seedling emergence by wheat residues collected 1 month after glyphosate desiccation was lost for the 14 μg/g soil N and 25/21°C day/night temperature treatment when compared to the 0 μg/g soil N and 25/21°C day/night temperature treatment (Lehman and Blum 1997). This effect of N was not observed at higher temperatures (i.e., 30/26 and 35/31°C), and
b. The inhibition of pigweed seedling emergence for the highest soil water level (0.173 g water/g soil) and temperature (30/26°C day/night) tested was reduced or completely eliminated in the presence of inhibitory wheat shoot residues (collected 2 months after desiccation) and non-inhibitory crimson clover residues (collected 4 month after desiccation), but not in the presence of inhibitory crimson clover shoot residues (collected 1 month after desiccation; Fig. 3.16; Lehman and Blum 1997). The reduction or elimination of inhibition was likely due to the binding of excess water by the cover crop residues. For the inhibitory crimson clover shoot residue we suspect the inhibitory activity of the residue was greater than the benefits derived from the ability of the residues to bind excess water.

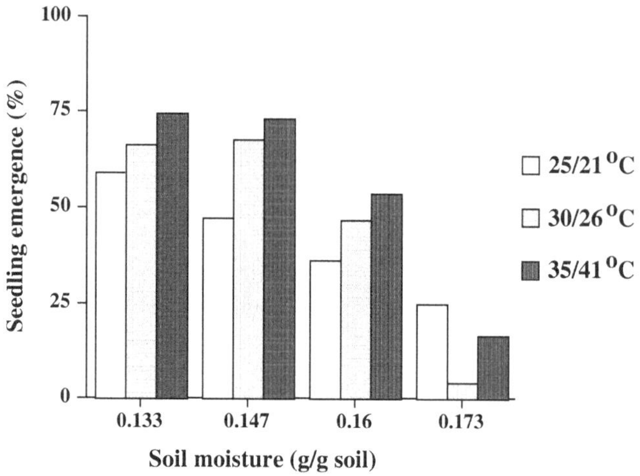

Fig. 3.15 The emergence of pigweed seedlings in Cecil A soil at 4 water levels and 3 day/night temperatures. Figure based on regressions (r^2 for 25/21°C = 0.57, for 30/26°C = 0.88, and for 35/41°C = 0.87) from Lehman and Blum (1997). Regressions used with permission of International Allelopathy Foundation

What about nutrient content within plant tissues/residues? Hall et al. (1983) found that total phenolic acid content of sunflower tissues mixed into soil did not account for a significant amount of inhibition of pigweed seedling biomass unless total N, P, and K content of the sunflower tissues were also included in their regression models. N was negatively related, and P and K were positively related to pigweed seedling dry weight biomass. The authors suggested that the effects of total N in sunflower tissues might have been due to:

a. inhibitors containing nitrogen,
b. nitrogen in a readily unavailable form that was closely related to total phenolic acid content (and/or other sunflower tissue inhibitors; see Macías et al. 1999),
c. rapid microbial utilization of nitrogen in sunflower tissues with an associated release of inhibitors or conversion of non-inhibitory substances to inhibitory substances, and/or
d. competition for N in sunflower tissues by pigweed seedlings and soil/or rhizosphere microbes.

Adding Hoagland's solution in addition to sunflower tissues eliminated the inhibitory effects of sunflower tissues on pigweed seedling biomass. Thus, it would appear that nutrients in plant tissues mixed into soil might modify plant–plant allelopathic interactions differently than nutrients/fertilizer directly applied to surface of the soil or mixed into the soil. The functional role of nutrients in small grain and clover cover crop residues is presently not known.

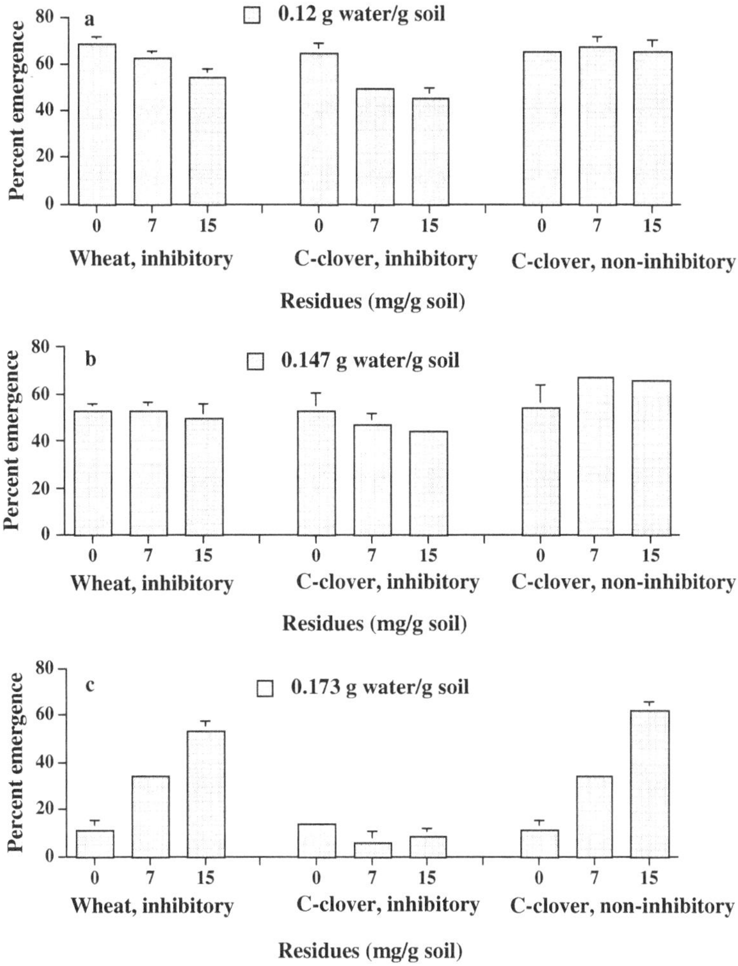

Fig. 3.16 The effects of soil moisture, and wheat and crimson clover cover crop residues on percent pigweed seedling emergence in Cecil A soil, where C equals crimson. Wheat inhibitory, C-clover inhibitory, and C-clover non-inhibitory were collected 2, 1, and 4 months after glyphosate desiccation, respectively. The absence of standard error bars indicates that the error bars are too *small* to be visible. Figures adapted from Lehman and Blum (1997). Figures used with permission of International Allelopathy Foundation

Finally, the potential role of other inhibitors should also be acknowledged. A variety of other inhibitors, besides phenolic acids, have been identified for the cover crop shoot tissues/residues, among them are organic acids, hydroxamic acids, and volatile compounds (Patrick 1971; Lynch 1977; Chou and Patrick 1976; Willard and Penner 1976; Tang and Waiss 1978; Buttery et al. 1985; Shilling et al. 1985; Niemeyer et al. 1989; Nair et al. 1990; Bradow and Connick 1990; Bradow 1991; Wu et al. 2001c; Macías et al. 2005; Mathiassen et al. 2006; also see Section 3.1).

In summary inhibitory effects on weed emergence by shoot cover crop tissues/residues mixed into soil can vary with weed species, species and cultivar of cover crop tissue/residue, length of time residue is left in the field after glyphosate desiccation, chemical composition of the tissue/residue, and/or the nature of the physicochemical environment of the soil. Since the addition of shoot cover crop tissues/residues on and in the soil are known to modify the biotic and physicochemical nature of the soil environment (e.g., change microbial activity, soil moisture, temperature, and nutrient content), shoot cover crops and cover crop tissues/residues could effectively inhibit weed seedling emergence by way of release of phenolic acids and/or other inhibitors as well as by modifying the soil environment. The resulting changes in the soil environment (e.g., microbial populations, moisture, temperature, and nitrogen) can also influence the expression of allelopathic interactions (see Section 2.6.1). In addition the effects of organic and inorganic compounds released from the cover crop tissues/residues can also be influenced by the presence of organic and inorganic components already in the soil. These observations suggest that because of the complex nature of soil environments, simple one to one relationships between soil inhibitor concentrations such as phenolic acids and weed seedling emergence are unlikely to be observed under field conditions except in very unusual circumstances.

3.4.6 Determine the Respective Importance of Shoot and Root Residues in Regulating Early Broadleaf Weed Seedling Emergence (Blum et al. (2002); Summarized with Permission of International Allelopathy Foundation)

Since for the 1992 and 1993 no-till field studies (Blum et al. 1997) both root and shoot residues of rye, wheat, crimson clover, and subterranean clover were present in each no-till plot, the role of shoot or root residues alone on weed seedling emergence could not be determined. Thus for the 1996 and 1997 field studies, sets of no-till plots with only root residue (i.e., shoots removed), only shoot residue (i.e., cut shoots from root plots placed on no-till reference plots), and shoot (uncut and cut but left in place) and root residues were included in the experimental design (Blum et al. 2002). However, to make the 1996 and 1997 field studies more manageable, only one cover crop, wheat, was utilized.

The maximum weed seedling emergence in the field for morningglory, pigweed and prickly sida ranged from < 1 to 23%. This was consistent with the earlier weed seedling emergence study (Blum et al. 1997). Only average weed seedling emergence for each experimental period will be presented here, for data on weed seedling emergence during the 1996 and 1997 experimental periods and for the different glyphosate-desiccation times in 1997 see Blum et al. (2002). Glyphosate desiccation of the wheat cover crop occurred on May 21, 1996 and April 21, April 30, and May 7, 1997.

In general wheat root residues alone stimulated mean morningglory and prickly sida seedling emergence and inhibited pigweed seedling emergence when compared to no-till reference plots without root residues (Figs. 3.17 and 3.18). Stimulations ranged from 152 to 382% and inhibition ranged from 84 to 96% for the mean seedling number per unit area over the experimental period (Blum et al. 2002). The presence of wheat shoot residues alone had no significant observable effect on seedling emergence of weed seedlings. The effects of shoot (shoot cut and uncut)

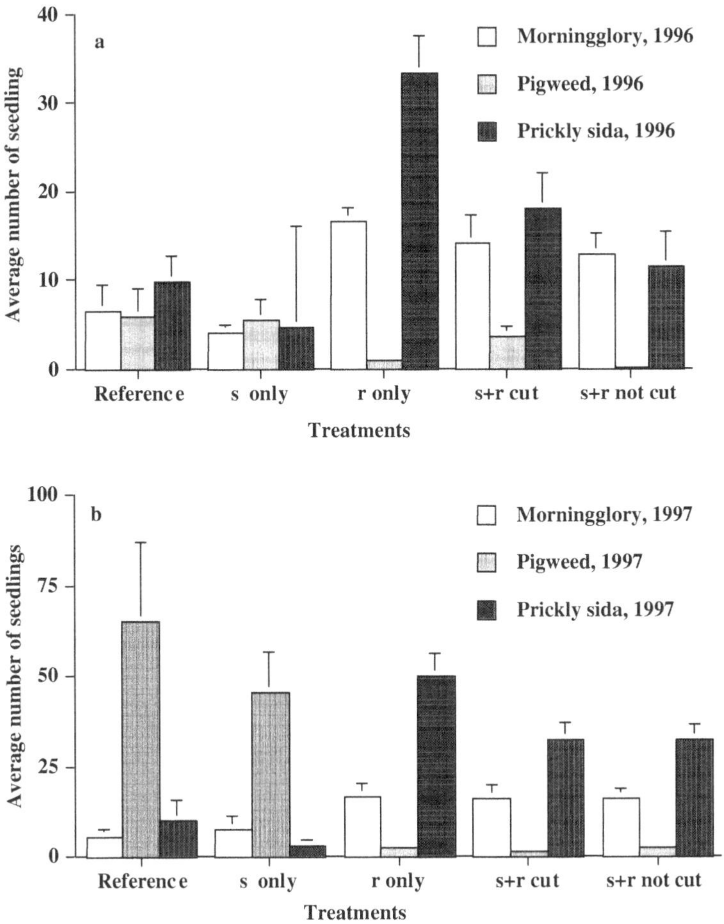

Fig. 3.17 Average number of morningglory, pigweed, and prickly sida seedlings in no-till Cecil A soil field plots for two experimental periods [(**a**) 1996 and (**b**) 1997] with the following 5 treatments: 1. no cover crop (reference), 2. cut wheat shoots on surface (s only), 3. wheat roots left in place but shoots cut and removed (r only), 4. wheat shoots and roots left in place, but shoots cut (s+r cut), and 5. wheat shoots and roots left in place, but shoots not cut (s+r not cut). The absence of standard error bars indicates that the error bars are too *small* to be visible. Figures based on data from Blum et al. (2002). Data used with permission of International Allelopathy Foundation

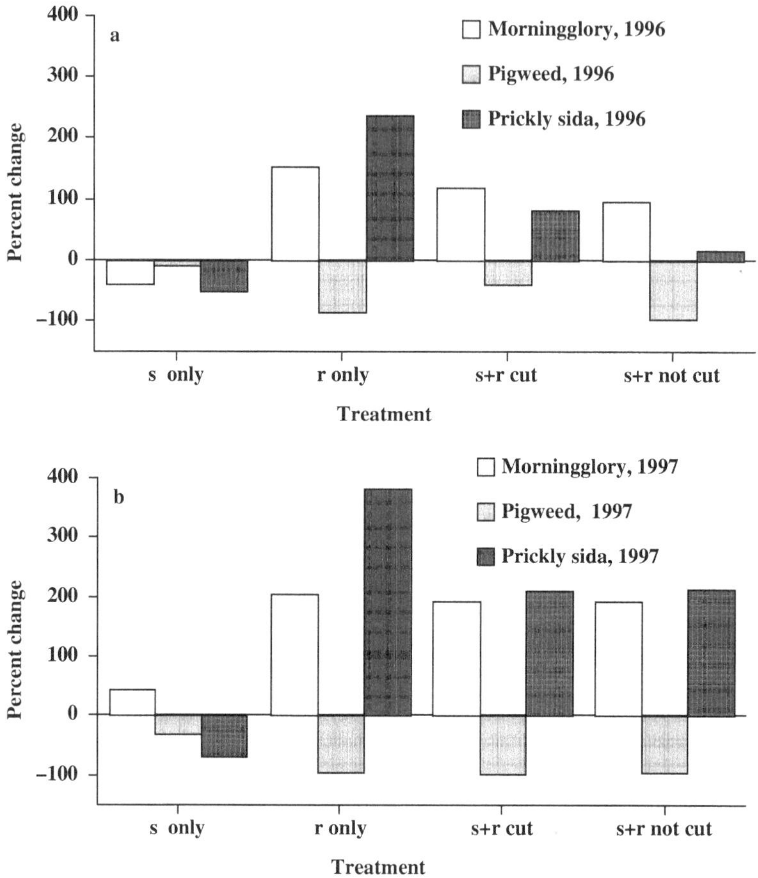

Fig. 3.18 Percent change of morningglory, pigweed, and prickly sida seedlings in no-till Cecil soil field plots for two experimental periods [(**a**) 1996 and (**b**) 1997] with the following 4 treatments: 1. cut wheat shoots on surface (s only), 2. wheat roots left in place but shoots cut and removed (r only), 3. wheat shoots and roots left in place, but shoots cut (s+r cut), and 4. wheat shoots and roots left in place, but shoots not cut (s+r not cut). Figures based on data from Fig. 3.17. Original data from Blum et al. (2002). Data used with permission of International Allelopathy Foundation

and root residue combinations were inconsistently expressed in 1996 (Fig. 3.18). In general for 1996 and 1997, when significantly different from the no-till reference plots, morningglory and prickly sida seedling emergence were stimulated and pigweed seedling emergence was inhibited. The effects of root residues were clearly evident even in the presence of shoot residues. Time of glyphosate desiccation in 1997 was significant for morningglory and pigweed seedling emergence but not for prickly sida. Numbers of morningglory and pigweed seedlings were highest for the first desiccation period (April 21) and the lowest for the second and third desiccation periods (April 30 and May 7). As in previous attempts, the use of environmental data (e.g., soil pH, soil moisture, and soil temperature) to establish relationships between the environmental data and observed morningglory, pigweed, and prickly sida seedling emergence were not successful.

Why the dramatic differences for seedling emergence between the two field studies (1992 and 1993 seasons; Blum et al. 1997 vs. 1996 and 1997 seasons; Blum et al. 2002)? Two differences immediately come to mind:

a. Different wheat cultivars were used for the two studies ("Coker 983" vs. "Southern States 555") because "Coker 983" used for the first study was no longer commercially available for the second study, and
b. Rainfall was below normal for all the experimental periods but moisture stress was more severe for the second study. For example total rainfall during periods when seedlings were counted each year was as follows: June to July 1992 – 21 cm (18% below normal), May to July 1993 – 19 cm (47% below normal) plus 2.54 cm of irrigation, June to July 1996 – 11 cm (57% below normal) and May to June 1997 – 12 cm (43% below normal). Reductions from normal were based on 1900–1962 mean rainfall data for Raleigh, NC, approximately five miles from the field site. However, soil moisture was adequate for weed seedling emergence and survival except for 1993 when plots were surface irrigated once (2.54 cm) in June.

In laboratory studies pigweed seedling emergence was inhibited more by "Southern States 555" shoot residues than root residues, and "Coker 983" shoot residues were more inhibitory than shoot residues of "Southern States 555" (Lehman and Blum 1997; Staman et al. 2001). For these studies wheat residues were ground and mixed into the soil in order to maximize potential effects of residues on pigweed seedling emergence (Blum 1999). For the no-till plots in the field, however, the wheat shoot residues were concentrated on the surface of the soil (i.e., not ground and mixed into soil) and the wheat root residues were distributed throughout the upper soil horizons (i.e., "diluted" but not ground and mixed into the soil). Surface area contact of wheat shoot residues and wheat root residues with soil were thus very different between the laboratory and field studies. Potential and actual transfer of organic and inorganic compounds from shoot and root residues to soil were thus also very different. Transfer of organic and inorganic compounds from shoot residues to the field soil were primarily limited by low surface area (i.e., little, if any, fragmentation) for leaching and lack of rainfall events of sufficient magnitude, duration, and frequency. Transfer from root residues was primarily limited by microbial/faunal activity and soil moisture. We suspect low rates and timing of rainfall events during seedling emergence and the absence of irrigation in the second study (irrigation occurred once during the 1993 study; Blum et al. 1997) may help to explain why wheat shoot residues for the second study (Blum et al. 2002) had little, if any, impact on weed seedling emergence. Unfortunately, since separate shoot and root residue treatments were not included in the first study, we do not know how shoot and root residues under those environmental conditions might have influenced weed seedling emergence.

Given the above scenarios, why did the incorporation by tilling (mixing) of living wheat shoot tissues into the Cecil A soil in 1993 lead to a stimulation of weed seedling emergence instead of an inhibition? After all both shoot and root tissues were now mixed into the soil resulting in greater surface area for the transfer of organic and inorganic compounds to the moist soil. Furthermore, living tissues have been found to be more inhibitory than aged residues in laboratory bioassays (Lehman and Blum 1997). Several potential reasons suggest themselves:

a. The zone of inhibition near the surface of the soils that is thought to develop over time under surface residues did not have time to develop (i.e., living standing wheat shoot tissues were incorporated into the soil) or if present, was disrupted by the wheat shoot tissues incorporation. Putman et al. (1983) suggested that sorghum surface residues produced a zone of inhibition near the surface of the soil where weed seeds germinate and that any disruption of this zone will eliminate the effects of this zone on weed seedling emergence.
b. Incorporation of living shoot tissue into the soil fragments both the shoot and root tissues, distributes tissues within a larger soil volume, increases the contact between soil and the cover crop tissues, and thus leads to lower available active organic compounds in the soil.
c. Roots in the soil were fragmented and mixed with the incorporated shoot tissues. Root effects may have dominated similar to that observed for the 1996/1997 experimental season or the promoter/modifier/inhibitor complex was shifted to the promoter side.
d. Fragmentation of living tissues during tillage was minimal and transfer of organic and inorganic compounds from living tissues is much slower than from dead tissues. Thus the rates and timing of release of agents required for inhibition were too low, too slow, and/or at the wrong time.
e. Tilling dramatically reduced the anaerobic site in the soil where phytotoxic substances accumulate, and
f. Tilling, the incorporation of wheat shoot tissues and the fragmentation and redistribution of root tissues, produced many new "safe" germination sites. Harper et al. (1965) found that increased variation in microtopography of soil surfaces was sufficient in itself to lead to greater weed seedling establishment.

In summary the effects of wheat shoot and/or root residues on weed seedling emergence in Cecil A soil varied considerably ranging from stimulation, to no effect, to inhibition depending on weed species, type of residues present, timing of desiccation, location of residues on or in the soil, and timing, frequency and amounts of rainfall. Organic and inorganic compounds in shoot wheat residues on the surface of the soil, however, will have little impact on weed seedling emergence unless environmental conditions (e.g., rainfall events) are appropriate for the release of these agents to the soil and the subsequent actions of these agents on weed seeds or roots. The excessive emphasis on the role of surface shoot residues in weed management may be misplaced since root residues may be just as or more important in determining weed seedling emergence.

3.4.7 Determine Under Controlled Conditions How Phenolic Acids-Containing Plant Tissues/Residues Mixed into Soil Modify Phenolic Acid-Utilizing Bulk-Soil and Rhizosphere Microbial Populations (Staman et al. (2001); Plenum Publishing Corporation, Excerpts Used with Permission of Springer Science and Business Media)

The addition of phenolic acids to soil can stimulate both bulk-soil and rhizosphere microbial populations that can utilize phenolic acids as a sole carbon source and/or inhibit the growth of cucumber seedlings (Blum and Shafer 1988; Shafer and Blum 1991; Blum et al. 1999a, b, 2000; Staman et al. 2001; Blum 2004, 2006). Blum et al. (2000) found that multiple additions of up to 0.6 μmol/g soil of an equal-molar phenolic acid mixture composed of *p*-coumaric acid, ferulic acid, *p*-hydroxybenzoic acid and vanillic acid resulted in a strong linear relationship between the inhibition of leaf expansion of cucumber seedlings growing in Cecil A soil (our model species; see Chapter 2) and increasing rhizosphere microbial populations that could utilize phenolic acids as a sole carbon source (Fig. 3.19). Additions beyond 0.6 μmol/g soil inhibited both seedling growth and phenolic acid-utilizing microbial populations. Since the utilization of phenolic acids by soil microbes generally leads to a reduction or loss of inhibition of seedling emergence and growth (Stowe and Osborn 1980; Sparling and Vaughan 1981; Sparling et al. 1981; Vaughan et al. 1983; Blum and Shafer 1988; Shafer and Blum 1991; Blum et al. 1987, 1999a, b, 2000), this linear relationship was somewhat counter intuitive and suggested that given these concentrations (up to 0.6 μmol/g) of phenolic acids and experimental and environmental conditions that inhibition of cucumber seedling growth and stimulation of microbial populations were, for all practical purposes,

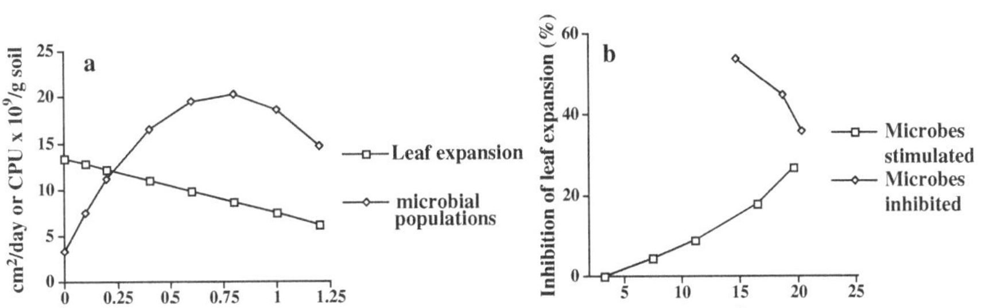

Fig. 3.19 Effects of total phenolic acid composed of a 4-equal-molar mixture of *p*-coumaric acid, ferulic acid, *p*-hydroxybenzoic acid, and vanillic acid on absolute rates of leaf expansion (cm^2/day; $r^2 = 0.44$) of 12 day-old cucumber seedlings and microbial populations (CFU/g soil; $r^2 = 0.49$) that can utilize phenolic acids as a sole carbon source in Cecil A soil (**a**). Relationships between phenolic acid-utilizing microbes (CFU, colony forming units) and percent inhibition of absolute rates of leaf expansion for cucumber seedlings are presented in (**b**). Values for (**b**) were calculated from (**a**). Figures based on regressions from Blum et al. (2000). Plenum Publishing Corporation, regressions used with permission of Springer Science and Business Media

independent. This independence of action could occur whenever the utilization of phenolic acid by bulk-soil and rhizosphere/rhizoplane microbes are limited because of insufficient numbers, uneven distribution (i.e., clumped or localized), and/or the physicochemical environment. It could also occur whenever transport of phenolic acids by gravitational flow, capillary action, or "transpirational pull" (mass flow) to seedling root surfaces is faster than can be utilized by microbes present (i.e, phenolic acid concentrations or rate of supply are sufficient to impact both microbes and roots).

Given that cover crop tissues/residues containing phenolic acids were known to inhibit a variety of seedlings (Hall et al. 1983; Blum et al. 1997, 2002; Lehman and Blum 1997), Staman et al. (2001) asked: Could the incorporation of plant tissues/residues containing phenolic acids such as wheat (Blum et al. 1992; Wu et al. 2000a, 2001b) or sunflower leaf tissue (Wilson and Rice 1968; Koeppe et al. 1969) simultaneously stimulate rhizosphere phenolic acid-utilizing microbes and inhibit the growth of seedlings? Although the focus was on phenolic acids clearly others inhibitors were also present in wheat shoot tissues/residues (Patrick 1971; Willard and Penner 1976; Lynch 1977; Tang and Waiss 1978; Buttery et al. 1985; Shilling et al. 1985; Niemeyer et al. 1989; Bradow and Connick 1990; Bradow 1991; Wu et al. 2001c) and sunflower leaf tissue (see Macías et al. 1999). Furthermore, if such direct relationships were observed, not only under laboratory conditions, but also under field conditions, this would provide clear but indirect evidence that phenolic acids were transferred from organic residues located in the bulk soil or just outside the rhizosphere to the rhizosphere/rhizoplane and that phenolic acids released from the residue were involved in plant–plant allelopathic interactions under field conditions.

In a set of preliminary experiments growth inhibition of cucumber seedlings and/or stimulation of phenolic acid-utilizing microorganisms in the rhizosphere of cucumber seedlings were observed when wheat ("Southern States 555") or sunflower tissues were mixed into soil (Figs. 3.20 and 3.21; Staman et al. 2001). Unfortunately inhibition of seedling growth and stimulation of phenolic acid-utilizing microbes in the rhizosphere were inconsistently observed among experiments and over time (Figs. 3.20 and 3.21; Staman et al. 2001). Field trials have so far not been initiated. However, given the right conditions (e.g., phenolic acid sensitive weed species, appropriate levels of phenolic acid concentrations in the soil residues, and appropriate phenolic acid-utilizing bulk-soil and rhizosphere microbial populations), such an approach may still prove to be useful in providing indirect evidence for the transfer and role of phenolic acid in plant–plant allelopathic interactions under field conditions. Such potential relationships should also be explored for other inhibitors in cover crop residues.

In summary soil incorporation of inhibitory concentrations of tissues, such as wheat and sunflower resulted in a simultaneous stimulation of rhizosphere microbial populations that could utilize phenolic acids as a sole carbon source and an inhibition of the growth of cucumber seedlings. However, this relationship was

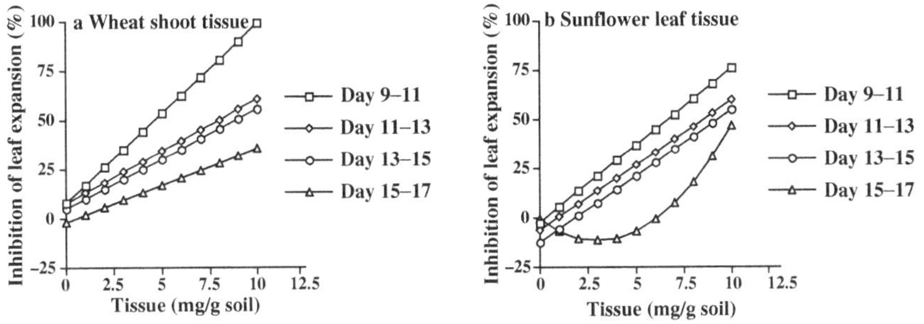

Fig. 3.20 Effects of wheat shoot (**a**; r^2 ranged from 0.54 to 0.80) and sunflower leaf (**b**; r^2 ranged from 0.55 to 0.77) tissues incorporated into Cecil A soil on percent inhibition of absolute rates of leaf expansion of cucumber seedlings over time. Figures based on regressions from Staman et al. (2001). Plenum Publishing Corporation, regressions used with permission of Springer Science and Business Media

inconsistently expressed among experiments and over time. Even with this limitation this experimental tool warrants further exploration, not only for phenolic acids, but also for a range of other plant–plant allelopathic inhibitors.

3.5 Summary of Major Points

3.5.1 Effects of Cover Crop Residues on the Physicochemical Environment of the Soil

a. The presence of glyphosate-desiccated cover crops in no-till plots reduced the solar radiation at the soil surface (for example > 85% for red and blue wavelengths in June of 1996), reduced the soil temperature (at 2.5 cm 4–10°C), and increased soil moisture (0–2.5 cm soil samples by 0.02–0.05 g water/g soil) when compared to no-till reference plots in 1992 and 1993.
b. Soil nitrate-N values for glyphosate-desiccated clover cover crops in no-till plots were higher by 1.6–6.2 μg/g soil in 1992 and by 4–32 μg/g soil in 1993 when compared to no-till cover crop plots. Changes for small grain cover crop plots ranged from plus 4 to minus 7 μg/g soil when compared to no-till cover crop plots. Small grain cover crop plots were fertilized with NH_4NO_3 in spring of 1993.
c. The presence of glyphosate-desiccated cover crops modified phenolic acid content of soil, for summary see Section 3.5.2, and
d. The presence of glyphosate-desiccated cover crops in no-till plots did not substantially modify soil pH, soil compaction, and C/N ratios of soil.

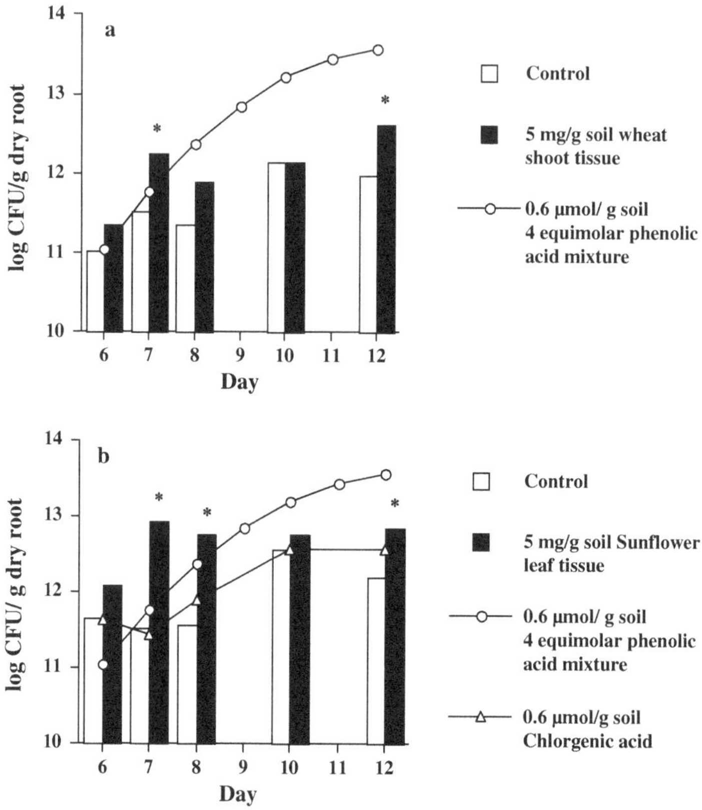

Fig. 3.21 Effects of wheat shoot (**a**) and sunflower leaf (**b**) tissues, a phenolic acid mixture composed of equal-molar concentrations of *p*-coumaric acid, ferulic acid, *p*-hydroxybenzoic acid, and vanillic acid (**a**, **b**; $r^2 = 0.83$), or chlorogenic acid (**b**) supplied to Cecil A soil on rhizosphere phenolic acid-utilizing microbes, where CFU equals colony forming units. The phenolic acid mixture and the chlorogenic acid were applied every other day to the soil while the shoot and leaf tissues were added to the soil only once, at the beginning of the experiment. Sunflower tissues and chlorogenic acid were incorporated and supplied, respectively, to a batch of autoclaved soil. This autoclaved soil, however, was not sterile. Soils were autoclaved only once to reduce the initial microbial populations. *Asterisks* indicate significant difference from the control (alpha = 0.05). Figures based on data and regressions from Staman et al. (2001). Plenum Publishing Corporation, data and regressions used with permission of Springer Science and Business Media

3.5.2 Phenolic Acids in Cecil Soils

a. Available phenolic acids in soils range from simple compounds, such as benzoic acids, to complex polymers, such as tannins. Soils also contain a variety of polymers that contain phenolic acid moieties, e.g., humic acid, fulvic acid, lignin, and melanins. Depending on the type of phenolic acid different extraction, identification, and quantification procedures are required.

b. Water, neutral EDTA and water-autoclave extractions in conjunction with HPLC analysis and/or the Folin & Ciocalteu's phenol reagent were found to be useful in recovering and quantifying available phenolic acids in Cecil soils. HPLC analysis was used to quantify individual phenolic acids and the Folin & Ciocalteu's phenol reagent was used to quantify total phenolic acid. Neutral EDTA extracts could not be used with the Folin & Ciocalteu's phenol reagent because EDTA reduces the phenol reagent.
c. Water extractions recovered "free" phenolic acids in soil solutions. Neutral EDTA extractions recovered "free" and reversibly sorbed phenolic acids in soils. Water-autoclave extractions recovered "free", reversibly sorbed, and some irreversibly sorbed phenolic acids in soils.
d. Water-autoclave extractions were much more effective than neutral EDTA in recovering phenolic acids from organic residues. Both water and neutral EDTA recovered little, if any, tightly sorbed (fixed) phenolic acids from organic residues.
e. Total phenolic acids in soils as determined by the Folin & Ciocalteu's phenol reagent represent relative, not absolute, values that can only be used for comparisons of similar types of soil and plant tissue/residue. This is because the Folin & Ciocalteu's phenol reagent reacts not only with phenolic acids but also some other organic and inorganic compounds and color development varies with phenolic acids (and other reducing organic and inorganic compounds) present.
f. Estimates of phenolic acid concentrations in soils represent static residual or net concentrations, i.e., what is left after inputs and losses to various soil sinks. Determinations made over time may provide some insight into net flows (in-out) but these can be misleading since flows occur even when no changes can be detected, i.e. input equals output. Soil extractions thus either tend to overestimate or underestimate the actual available phenolic acids in soils depending on circumstances.
g. Phenolic acid concentrations in cover crop residues were considerably higher than in the soil. For example the concentration of *p*-coumaric acid was 65 times higher in wheat straw from half buried litter bags than in 0–2.5 cm soil cores.
h. Readily available phenolic acids in desiccated cover crops (e.g., crimson clover, subterranean clover, wheat, and rye) on or near the surface of the soil declined after glyphosate desiccation. Observed phenolic acid losses/retentions from residues varied with species and appeared to be related to frequency, magnitude, and length of rainfall events. We would thus anticipate that input of phenolic acids into the soil from cover crop residues would parallel the decline of phenolic acids in the cover crop residues. Note: We would also expect similar behavior for other available organic and inorganic compounds.
i. One major source of available/active phenolic acids in no-till systems was the plant residue on and in the soil. In general, highest concentrations of phenolic acids occur at or near the soil surface due to the leaching of the compressed biomass of the shoot cover crop residues located there. Phenolic acid released from root residues, on the other hand, have a much larger vertical distribution in the soil. Phenolic acids decline with soil depth.

j. Highest average phenolic acid concentration recovered from Cecil A soil samples was 4 μg/g soil for *p*-coumaric acid, a cinnamic acid. In general benzoic acids (vanillic, *p*-hydroxybenzoic acid, and sinapic acid) were higher in concentration than cinnamic acids (caffeic acid, ferulic acid, and syringic acid), with the exception of *p*-coumaric acid.
k. Individual, the sum of 7 common phenolic acids, and total phenolic acid content in soil extracts from Cecil A soil were found to be correlated with each other and with moisture, pH, carbon, and nitrogen of the soil.
l. Total phenolic acid content of wheat no-till, wheat conventional-till, and fallow conventional-till soil samples was approximately 15 times higher than the sum of caffeic acid, *p*-coumaric acid, ferulic acid, *p*-hydroxybenzoic acid, sinapic acid, syringic acid, and vanillic acid. Concentrations for the sum of the 7 phenolic acids were 34% greater in 0–2.5 cm soil samples than in 0–10 cm soil samples.
m. Concentrations of individual, sum of 7 phenolic acids, and total phenolic acid content in Cecil soil appeared to be fairly constant over the experimental periods studied. For example, the maximum changes for the sum of 7 phenolic acids over the experimental periods were on the order of 20–30%.
n. Soil concentrations for the sum of 7 phenolic acids (0–2.5 cm) were 42 and 62% higher for wheat no-till soils (12 μg/g) than for wheat conventional-till soils and fallow conventional-till soil samples, respectively, and
o. Total phenolic acid content in no-till soils (0–2.5 cm) were equal to or greater for glyphosate-desiccated clover and small grain cover crop soils (mean of 73 μg/g soil in 1993) than for no-till reference soils (mean of 62 μg/g soil in 1993). Soil concentrations were higher for clover cover crop soils (mean of 83 μg/g soil in 1993) than for small grain cover crop soils (mean of 63 μg/g soil in 1993). Note: Total phenolic acid content of small grain and clover residues and thus soils associated with them should probably not be compared directly because the Folin & Ciocalteu's phenol reagent is also reduced by a number of other organic and inorganic compounds, including several amino acids that are higher in clover residues.

3.5.3 Bioassays of Soil Extracts

a. Germination and radicle/hypocotyl growth of weed seeds are affected by many physical and chemical characteristics of soil extracts such as presence, concentrations, and activities of promoters, modifiers, and inhibiters, pH, solute potential, and organic and inorganic (e.g., nutrients) compounds. Effects of soil extracts on seed germination and radicle/hypocotyl growth are thus not the exclusive property of allelopathic agents in the extract but a function of all such factors acting independently and/or in a synergistic, additive, or antagonistic manner.
b. Since physicochemical properties (e.g., pH, nutrients, organic particles/residues, and phenolic acid content) of soils are frequently uneven in distribution in

field plots and at times co-linear, obtaining soil extracts that have the necessary and independent ranges of variables of interest for bioassays and regression or response surface analyses is difficult even under the best of circumstances. The identification of inhibitory or stimulatory soil properties are further confounded by experimental noise generated by soil extracts which are borderline or inactive (i.e., not inhibitory or stimulatory) to the processes being measured in the bioassay. Thus using soil extracts directly in bioassays may limit our ability to identify and sort out inhibitory or stimulatory soil extract properties.

c. Water-autoclave extracts of ± wheat conventional-tilled soils did not inhibit germination and radicle/hypocotyl lengths of crimson clover and morningglory. Extracts of wheat no-till soils also did not inhibit germination but inhibited radicle/hypocotyl lengths of both species. However, there appeared to be a co-linearity for the inhibitory effects of total phenolic acid and solute potential (see b.).
d. Biological activity of individual extracts turned out to be much more helpful in identifying the inhibitory properties of extract solutions than using soil extracts directly (see b and c). Serial dilutions of each extract, i. e., dose response, were thus used to determine the inhibition slopes for each extract. Linear slopes with a common intercept were found to be adequate to describe the slopes. Significant slopes were used as proxies for biological activity (Blum et al. 1992). Biological activities, i.e., inhibition slopes of extracts, were then regressed against initial undiluted extract properties (e.g., solute potential, pH, total phenolic acid, and nutrients) to identify active properties of soil extract solutions, and
e. For adequately sampled soil plots the frequency of significant extract slopes can be used as an indicator of the potential importance of allelopathic interactions in those plots.

3.5.4 Field Residue Bioassays: Seedling Emergence

3.5.4.1 1992 and 1993 Experimental Periods

a. Maximum weed seedling emergence in the field plots ranged from < 1 to 23% for the broadcast seeds. Laboratory germination at time of seeding ranged from 22 to 70%. Thus a large fraction of seeds broadcast into the subplots were either inactive (quiescent or dormant) or lost because of predation and/or mortality.
b. Mean morningglory, pigweed, and prickly sida seedling emergence for the entire experimental period in glyphosate-desiccated crimson clover, subterranean clover, rye, and wheat shoot and root residue plots varied with year, time of glyphosate desiccation, cover crop species, and weed species when compared to no-cover crop reference plots. Responses ranged from stimulation (mean maximum 76%) to inhibition (mean maximum 96%).
c. All four glyphosate-desiccated cover crop residues delayed the initial seedling emergence of morningglory, pigweed, and prickly sida compared to no-cover

crop reference plots. This was particularly evident in 1993. Once emergence was initiated, however, the rates of seedling emergence varied with time of glyphosate desiccation, cover crop species, and weed species.

d. Desiccated rye and wheat cover crops were the most consistent in the inhibition of weed seedling emergence.
e. Weed seedling emergence was stimulated (mean maximum 409%) when living cover crop biomass was incorporated into the soil, and
f. Cover crop residues provided only very weak control of broadleaf signal grass suggesting that control of monocotyledonous species may be a serious problem.

3.5.4.2 1996 and 1997 Experimental Periods

a. Wheat shoot residue had essentially no effect on weed seedling emergence of morningglory, pigweed, and prickly sida. Lack of inhibitory effects of wheat shoots for this study may have been related to the insufficient rainfall events during this study.
b. Wheat root residues significantly stimulated (mean maximum 382%) mean morningglory and prickly sida seedling emergence and significantly inhibited (mean maximum 96%) pigweed seedling emergence, and
c. The excessive emphasis on the role of surface shoot residues in weed management may be misplaced since root residues may be just as or more important in determining weed seedling emergence.

3.5.4.3 All Experimental Periods

a. No significant relationships between seedling emergence and measured bulk-soil physicochemical factors (soil total phenolic acid content, nitrate, pH, moisture, and temperature) were identified. Likely reasons for the lack of significant relationships were:

   1. bulk-soil total phenolic acid content was not a good proxy for promoters/modifiers/inhibitors complex,
   2. the action of promoters/modifiers/inhibitors occurs within the rhizosphere/rhizoplane and not in the bulk soil, and
   3. the ranges of environmental variation generated by the various cover crops in combination with the limited environmental data collected for the bulk soil were not sufficiently representative of the seed and root environments.

b. Indirect evidence suggested that timing, frequency, and length of rainfall events right after cover crop desiccation were important contributors in determining the rate and the magnitude of stimulation/inhibition for weed seedling emergence.
c. There were no clear cut and consistent patterns between biomass of cover crop tissues/residues, phenolic acid content of cover crop tissues/residues, and available phenolic acid content of bulk soils. This is not entirely surprising

since the readily available phenolic acids ("free" and reversibly sorbed) in cover crops are leached by rainfall/irrigation events shortly after glyphosate desiccation and the release (rate and timing) of the remaining tightly bound phenolic acids are regulated by decay and subsequent rainfall/irrigation events, a slow and extended process. In addition, once in the soil phenolic acids are sorbed, utilized by microbes, taken up by roots, transferred horizontally and/or vertically by gravitational flow, capillary action, and "transpirational pull". All of these factors and processes are highly variable in time and space, and

d. The inconsistent results, ranging from stimulation to inhibition, for weed seedling emergence and phenolic acid content of soil suggested that phenolic acids might not be an important regulating factor or that other factors (e.g., the physiochemical and biotic environments, and other organic and inorganic compounds) were modulating/regulating the effects of phenolic acids, i.e., phenolic acids were part of a promoter/modifier/inhibitor complex (see Section 2.6.1).

3.5.5 Laboratory Bioassays: Seedlings and Microbes

3.5.5.1 Seedling Emergence

a. Inhibitory effects of shoot residues on weed seedling emergence varied with weed species, species of cover crop residues, age of the residue in the field after glyphosate desiccation, and the nature of the physicochemical environment of the soil.
b. All freeze-dried ground shoot tissues of crimson clover, subterranean clover, rye, and wheat collected from the field prior to glyphosate desiccation when ground and mixed into soil at an appropriate concentration inhibited seedling emergence of morningglory and pigweed. One month after glyphosate desiccation only crimson clover and wheat shoot residues were inhibitory. More than 1 month after glyphosate desiccation only wheat shoot residues were inhibitory to pigweed seedling emergence.
c. Total phenolic acid content of field cover crop tissues/residues after glyphosate ranged from 14 mg/g for wheat to 33 mg/g for subterranean clover. After 1 month total phenolic acid content of tissues/residues was not changed for rye (+3%), increased for wheat (+37%), and declined for crimson clover (–45%) and for subterranean clover (–48%). At 2 months the decline of total phenolic acid ranged from 7% for wheat to 57% for subterranean clover. Note: Total phenolic acid content of small grain and clover residues should not be compared directly because the Folin & Ciocalteu's phenol reagent is reduced by a number of other organic and inorganic compounds, including several amino acids that are higher in clover residues.
d. Total phenolic acid content of cover crop tissues/residues was significantly related to the biological activity (i.e., inhibition slopes) of morningglory seedling emergence ($r^2 = 0.48$) but not pigweed seedling emergence.

e. Actions and/or interactions of soil moisture, soil pH, soil N, and/or growth temperature with cover crops tissues/residues at times modified, reduced, and/or eliminated the inhibitory effects of cover crop tissue/residues on morningglory and/or pigweed seedling emergence, and
f. Nutrient content of the plant tissues/residues and nutrients (e. g., Hoagland's) supplied directly to soil appeared to influence the inhibitory effects of plant tissues/residues differently.

The last three statements (i.e., e., f., and g.) are all consistent with the observation in Section 3.5.4.3 d. that other factors (e.g., the physiochemical and biotic environments and other organic and inorganic compounds) modulate/regulate the effects of phenolic acids. Additional support for the role of physicochemical and biotic environmental factors in determining the effects of phenolic acids was also provided by our model system studies with phenolic acids (see Section 2.5).

3.5.5.2 Simultaneous Effects on Seedlings and Microbes

a. Addition of freeze-dried ground wheat or sunflower tissues to soil stimulated bulk-soil and rhizosphere microbes that could utilize phenolic acids as a sole carbon source and/or inhibited leaf area expansion of cucumber seedlings, our model species.
b. Simultaneous stimulation of phenolic acid-utilizing microbes and inhibition of cucumber leaf expansion were observed for wheat or sunflower tissues, however, the simultaneous stimulation and inhibition was inconsistently expressed among experiments and over time.
c. When both occur the simultaneously effects of wheat and sunflower tissues could be driven either by phenolic acids or some combination of phenolic acids in conjunction with other promoters, modifiers, and/or inhibitors. Two possible scenarios:

   1. If both are driven by phenolic acids from the tissues, then the simultaneous effects could occur whenever transport of phenolic acids by gravitational flow, capillary action, or "transpirational pull" (mass flow) to seedling root surfaces is faster than can be utilized by microbes present (i.e, phenolic acid concentrations or rate of supply are sufficient to impact both microbes and roots independently).
   2. However, if phenolic acid from tissues are primarily utilized ("used up") to stimulate phenolic-acid utilizing microbes within the bulk soil and/or the rhizosphere/rhizoplane, then the inhibition of cucumber seedlings could be due to a promoter/modifier/inhibitor complex dominated by other inhibitors and any remaining phenolic acids and phenolic acid breakdown products, and

d. The relationships between rhizosphere/rhizoplane microbes and seedling growth should be explored further since it may potentially provide indirect evidence for the transfer of phenolic acid or other inhibitors from organic tissues/residues in soil to the rhizosphere/rhizoplane.

References

Anderson RA, Todd JR (1968) Estimation of total tobacco plant phenols by their bonding to polyvinylpyrrolidone. Tob Sci 12:107–111

Barnes JP, Putman AR (1983) Rye residues contribute weed suppression in no-tillage cropping systems. J Chem Ecol 9:1045–1057

Barnes JP, Putnam AR, Burke A (1986) Allelopathic activity of rye (*Secale cereale* L.). In: Putnam AR, Tang CS (eds) Science of allelopathy. Wiley, New York, NY, pp 271–286

Bates-Smith EC (1956) The commoner phenolic constituents of plants and their systematic distribution. Proc R Dublin Sci 27:165–176

Belz RG (2007) Allelopathy in crop/weed interactions – an update. Pest Manag Sci 63:308–326

Belz RG (2008) Stimulation versus inhibition – bioactivity of parthenin, a phytochemical from *Parthenium hysterophorus* L. Int Dose-Response Soc 6:80–96

Belz RG, Velini ED, Duke SO (2007) Dose/response relationships in allelopathy research. In: Fujii Y, Hiradate S (eds) Allelopathy: new concepts and methodologies. Science Publishers, Enfield, NH, pp 3–29

Bhowmik PC, Doll JD (1983) Growth analysis of corn and soybean response to allelopathic effects of weed residues at various temperatures and photosynthetic flux densities. J Chem Ecol 9:1263–1280

Blum U (1996) Allelopathic interactions involving phenolic acids. J Nematol 28:259–267

Blum U (1997) The benefits of citrate over EDTA for extracting phenolic acids from soils and plant debris. J Chem Ecol 23:347–362

Blum U (1998) Effects of microbial utilization of phenolic acids and their phenolic acid breakdown products on allelopathic interactions. J Chem Ecol 24:685–708

Blum U (1999) Designing laboratory plant debris-soil bioassays: some reflections. In: Inderjit, Dakshini KMM, Foy CL (eds) Principles and practices in plant ecology: allelochemical interactions. CRC Press, Boca Raton, FL, pp 17–23

Blum U (2004) Fate of phenolic allelochemicals in soils: the role of soil and rhizosphere microorganisms. In: Macías FA, Galindo JCG, Molinillo JMG, Cutler HG (eds) Allelopathy: chemistry & modes of action of allelochemicals. CRC Press, Boca Raton, FL, pp 57–76

Blum U (2006) Allelopathy: a soil system perspective. In: Reigosa MJ, Pedrol N, González L (eds) Allelopathy. A physiological process with ecological implications. Springer, Dordrecht, pp 299–340

Blum U (2007) Can data derived from field and laboratory bioassays establish the existence of allelopathic interactions in nature? In: Fujii Y, Hiradate S (eds) Allelopathy: new concepts and methodology. Science Publishers, Enfield, NH, pp 31–38

Blum U, Austin MF, Shafer SR (1999a) The fates and effects of phenolic acids in a plant-microbe-soil model system. In: Macías FA, Galindo JCG, Molinillo JMG, Cutler HG (eds) Recent advances in allelopathy I. A science for the future. Cádiz University Press, Puerto Real, pp 159–166

Blum U, Dalton BR, Shann JR (1985a) Effects of various mixtures of ferulic acid and some of its microbial metabolic products on cucumber leaf expansion and dry matter in nutrient culture. J Chem Ecol 11:619–641

Blum U, Dalton BR, Shann JR (1985b) Effects of ferulic and *p*-coumaric acids in nutrient culture on cucumber leaf expansion as influenced by pH. J Chem Ecol 11:1567–1582

Blum U, Gerig TM (2005) Relationships between phenolic acid concentrations, transpiration, water utilization, leaf area expansion, and uptake of phenolic acids: nutrient culture studies. J Chem Ecol 31:1907–1932

Blum U, Gerig TM (2006) Interrelationships between *p*-coumaric acid, evapotranspiration, soil water content, and leaf expansion. J Chem Ecol 32:1817–1834

Blum U, Gerig TM, Weed SB (1989) Effects of mixtures of phenolic acids on leaf area expansion of cucumber seedlings grown in different pH Portsmouth A1 soil materials. J Chem Ecol 15: 2413–2423

Blum U, Gerig TM, Worsham AD, Holappa LD, King LD (1992) Allelopathic activity in wheat-conventional and wheat-no-till soils: development of soil extract bioassays. J Chem Ecol 18:2191–2221

Blum U, Gerig TM, Worsham AD, King LD (1993) Modification of allelopathic effects of *p*-coumaric acid on morning-glory seedling biomass by glucose, methionine, and nitrate. J Chem Ecol 19:2791–2811

Blum U, King LD, Brownie C (2002) Effects of wheat residues on dicotyledonous weed emergence in a simulated no-till system. Allelopathy J 9:159–176

Blum U, King LD, Gerig TM, Lehman ME, Worsham AD (1997) Effects of clover and small grain cover crops and tillage techniques on seedling emergence of some dicotyledonous weed species. Am J Altern Agric 12:146–161

Blum U, Shafer SR (1988) Microbial populations and phenolic acids in soils. Soil Biol Biochem 20:793–800

Blum U, Shafer SR, Lehman ME (1999b) Evidence for inhibitory allelopathic interactions involving phenolic acids in field soils: concepts vs. an experimental model. Crit Rev Plant Sci 18:673–693

Blum U, Staman KL, Flint LJ, Shafer SR (2000) Induction and/or selection of phenolic acids-utilizing bulk-soil and rhizosphere bacteria and their influence on phenolic acid phytotoxicity. J Chem Ecol 26:2059–2078

Blum U, Weed SB, Dalton BR (1987) Influence of various soil factors on the effects of ferulic acid on leaf expansion of cucumber seedlings. Plant Soil 98:111–130

Blum U, Wentworth TR, Klein K, Worsham AD, King LD, Gerig TM, Lyu S-W (1991) Phenolic acid content of soils from wheat-no till, wheat-conventional till, and fallow-conventional till soybean cropping systems. J Chem Ecol 17:1045–1068

Blum U, Worsham AD, King LD, Gerig TM (1994) Use of water and EDTA extractions to estimate available (free and reversibly bound) phenolic acids in Cecil soils. J Chem Ecol 20: 341–359

Bonanomi G, Sicurezza MG, Caporaso S, Esposito A, Mazzoleni S (2006) Phytotoxicity dynamics of decaying plant materials. New Phytologist 169:571–578

Box JD (1983) Investigation of the Folin-Ciocalteau phenol reagent for the determination of polyphenolic substances in natural waters. Water Res 17:511–525

Bradow JM (1991) Relationships between chemical structure and inhibitory activity of C_6 through C_9 volatiles emitted by plant residues. J Chem Ecol 17:2193–2212

Bradow JM, Connick WJ (1990) Volatile seed germination inhibitors from plant residues. J Chem Ecol 16:645–666

Brouwer R (1983) Functional equilibrium: sense or nonsense? Neth J Agric Sci 31:335–348

Brust GE (1994) Seed-predators reduced broadleaf weed growth and competitive ability. Agric Ecosyst and Environ 48:27–34

Brust GE, House GJ (1988) Weed seed destruction by arthropods and rodents in low-input soybean agroecosystems. Am J Altern Agric 3:19–25

Buttery RG, Xu C-J, Ling LC (1985) Volatile components of wheat leaves (and stems): Possible insect attractants. J Agric Food Chem 33:115–117

Calabrese EJ, Baldwin LA (2002) Defining hormesis. Human Exp Toxicol 21:91–97

Cataldo DA, Haroom M, Schrader LE, Youngs VL (1975) Rapid colorimetric determination of nitrate in plant tissue by nitration of salicylic acid. Commun Soil Sci Plant Anal 6:71–80

Chou CH, Patrick ZA (1976) Identification and phytotoxic activity of compounds produced during decomposition of corn and rye residues in soil. J Chem Ecol 2:369–387

Dalton BR, Blum U, Weed SB (1983) Allelopathic substances in ecosystems: effectiveness of sterile soil components in altering recovery of ferulic acid. J Chem Ecol 9:1185–1201

Dalton BR, Weed SB, Blum U (1987) Plant phenolic acids in soils: a comparison of extraction procedures. Soil Sci Soc Am J 51:1515–1521

Davidson RL (1969a) Effects of root/leaf temperature differentials on root/shoot ratios in some pasture grasses and clover. Ann Bot 33:561–569

Davidson RL (1969b) Effects of soil nutrients and moisture on root/shoot ratios in *Lolium perenne* L. and *Trifolium repens* L. Ann Bot 33:571–577
DeFrank J, Putnam AR (1978) Weed and crop response to allelopathic crop residues. North Cent Weed Control Conf Proc 33:44
Duke SO, Cedergreen N, Velini ED, Belz RG (2006) Hormesis: is it an important factor in herbicide use and allelopathy? Outlook Pest Manag 2006:29–33
Einhellig FA (1987) Interactions among allelochemicals and other stress factors of the plant environment. In: Waller GR (ed) Allelochemicals: role in agriculture and forestry. ACS symposium series, vol 330. American Chemical Society, Washington, DC, pp 343–357
Einhellig FA, Leather GR (1988) Potentials for exploiting allelopathy to enhance crop production. J Chem Ecol 14:1829–1844
Elliott LF, Cheng HH (1987) Assessment of allelopathy among microbes and plants. In: Waller GR (ed) Allelochemicals: role in agriculture and forestry. ACS symposium series, vol 330. American Chemical Society, Washington, DC, pp 504–515
Enache A, Ilnicki RD (1988) Subterranean clover: a new approach to weed control. Proc Northeast Weed Sci Soc 42:34
Fester CR, Peterson GA (1979) Effects of no-tillage fallow as compared to conventional tillage in a wheat-fallow system. Agric Exp Sta Res Bull 289, University of Nebraska, Lincoln, NE
Flaig W (1971) Organic compounds in soil. Soil Sci 111:19–33
Gerig TM, Blum U (1991) Effects of mixtures of four phenolic acids on leaf area expansion of cucumber seedlings grown in Portsmouth B1 soil materials. J Chem Ecol 17:29–40
Glass ADM (1976) The allelopathic potential of phenolic acids associated with the rhizosphere of *Pteridium aquilinum*. Can J Bot 54:2440–2444
Goodwin TW, Mercer EI (1983) Introduction to plant biochemistry, 2nd edn. Pergaman Press, Oxford
Grabber JH (2005) How do lignin composition, structure, and cross-linking affect degradability? A review if cell wall model studies. Crop Sci 45:820–831
Guenzi WD, McCalla TM (1966) Phenolic acids in oats, wheat, sorghum, and corn residues and their phytotoxicity. Agron J 58:303–304
Hadas A, Kautsky L, Goek M, Kara EE (2004) Rates of decomposition of plant residues and available nitrogen in soil, related to residue composition through simulation of carbon and nitrogen turnover. Soil Biol Biochem 36:255–266
Hall AB, Blum U, Fites RC (1982) Stress modification of allelopathy of *Helianthus annuus* L. debris on seed germination. Am J Bot 69:776–783
Hall AB, Blum U, Fites RC (1983) Stress modification of allelopathy of *Helianthus annuus* L. debris on seedling biomass production of *Amaranthus retroflexus* L. J Chem Ecol 9: 1213–1222
Harborne JB (1990) Plant phenolics. In: Bell EA, Charlwood BV (eds) Secondary plant products. Springer, Berlin, pp 331–401
Harborne JB (1993) Introduction to ecological biochemistry, 4th edn. Academic Press Limited, London
Harborne JB (1998) Phytochemical methods: a guide to modern techniques of plant analysis, 3rd edn. Chapman & Hall, London
Harmsen GW, Van Schreven DA (1955) Mineralization of organic nitrogen in soil. Adv Agron 7:299–395
Harper JR, Balke NE (1981) Characterization of the inhibition of K^+ absorption in oat roots by salicylic acid. Plant Physiol 68:1349–1553
Harper JL, Williams JT, Sagar GR (1965) The behavior of seeds in soil I. The heterogeneity of soil surfaces and its role in determining the establishment of plants from seed. J Ecol 53:273–286
Hartley RD, Whitehead DC (1985) Phenolic acids in soils and their influence on plant growth and soil microbial processes. In: Vaughan D, Malcolm RE (eds) Soil organic matter and biological activity. Developments in plant and soil, vol 16. Martinus Nijhoff/Dr. W. Junk Publishers, Dordrecht, pp 109–150

Inderjit, Mallik AU (2002) Chemical ecology of plants: allelopathy in aquatic and terrestrial ecosystems. Birkäuser, Basel
Iritani WM, Arnold CY (1960) Nitrogen release of vegetable crop residues during incubation as related to their chemical composition. Soil Sci 89:74–82
Khanbabaee K, van Ree T (2001) Tannins: classification and definition. Nat Prod Rep 18:641–649
Kimber RWL (1973) Phytotoxicity from plant residues III. The relative effects of toxins and nitrogen immobilization on the germination and growth of wheat. Plant Soil 38:543–555
Koeppe DE, Rohrbaugh LM, Wender SH (1969) The effects of varying UV intensities on the concentration of scopolin and caffeoylquinic acids in tobacco and sunflower. Phytochem 8: 889–896
Lang ARG (1967) Osmotic coefficients and water potential of sodium chloride solutions from 0 to 40°C. Aust J Chem 20:2017–2013
Lavelle P, Spain AV (2001) Soil ecology. Kluwer Academic Publishers, Dordrecht
Lehman ME (1993) Effects of allelopathy on plant emergence and growth as modified by physical factors and root distribution. North Carolina State University Thesis, Raleigh
Lehman ME, Blum U (1997) Cover crop debris effects on weed emergence as modified by environmental factors. Allelopathy J 4:69–88
Lehman ME, Blum U (1999) Evaluation of ferulic acid uptake as a measurement of allelochemical dose: effective concentration. J Chem Ecol 25:2585–2600
Li X, Kellaway RC, Ison RL, Annison G (1992) Chemical composition and nutritive value of mature annual legumes for sheep. Animal Feed Sci Tech 37:221–231
Liebl RA, Worsham AD (1983) Inhibition of morning-glory (*Ipomoea lacunosa* L.) and certain other weed species by phytotoxic components of wheat (*Triticum aestivum* L.) straw. J Chem Ecol 9:1027–1043
Lynch JM (1977) Phytotoxicity of acetic acid produced in the anaerobic decomposition of wheat straw. J Appl Bact 42:81–87
Macías FA, Oliva RM, Varela RM, Torres A, Molinillo MG (1999) Allelochemicals from sunflower leaves cv. Peredovick Phytochem 52:613–621
Macías FA, Oliveros-Bastidas A, Marin D, Castellano D, Simonet AM, Molinillo JMG (2005) Degradation studies on benzoxazinoids. Soil degradation dynamics of (2R)-2-O-ß-D-glucopyranosyl-4-hydroxy-(2H)-1,4-benzoxazin-3-(4H)-one (DIBOA-Glc) and its degradation products, phytotoxic allelochemicals from gramineae. J Agric Food Chem 53:554–561
Mathiassen SK, Kudsk P, Mogensen BB (2006) Herbicidal effects of soil-incorporated wheat. J Agric Food Chem 54:1058–1063
McAllister RA (1969) Observations on the Folin-Ciocalteu reaction. J Med Lab Technol 26:1–10
Morrison FB (1956) Feeds and feeding: a handbook for students and stockman, 22nd edn. Morrison Publication Co, Ithaca, NY
Nagabhushana GG, Worsham AD, Yenish JP (2001) Allelopathic cover crops to reduce herbicide use in sustainable agricultural systems. Allelopathy J 8:133–146
Nair MG, Whitenack CJ, Putnam AR (1990) 2,2′-oxo-1,1′-azobenzene a microbially transformed allelochemical from 2,3-benzoazolione: I. J Chem Ecol 16:353–364
Nakano H, Morita S, Shigemori H, Hasegawa K (2006) Plant growth inhibitory compounds from aqueous leachates of wheat straw. Plant Growth Regul 48:215–219
Niemeyer HM, Pesel ES, Capaja SV, Bravo HR, Franke S, Francke W (1989) Changes in hydroxamic acid levels of wheat plants induced by aphid feeding. Phytochem 28:447–449
Ohno T, First PR (1998) Assessment of the Folin and Ciocalteu's method for determining soil phenolic carbon. J Environ Qual 27:776–782
Pandey DK (1994) Inhibition of salvinia (*Salvinia molesta* Mitchell) by parthenium (*Parthenium hysterophorus* L.). II. Relative effect of flower, leaf, stem, and root residue on salvinia and paddy. J Chem Ecol 20:3123–3131
Patrick ZA (1971) Phytotoxic substances associated with the decomposition in the soil of plant residue. Soil Sci 111:13–18
Paul EA, Clark FE (1989) Soil microbiology and biochemistry. Academic Press, San Diego, CA

Pue KJ, Blum U, Gerig TM, Shafer SR (1995) Mechanism by which noninhibitory concentrations of glucose increase inhibitory activity of *p*-coumaric acid on morning-glory seedling biomass accumulation. J Chem Ecol 21:833–847

Putnam AR, DeFrank J (1983) Use of phytotoxic plant residues for selective weed control. Crop Prot 2:173–181

Putnam AR, DeFrank J, Barnes JP (1983) Exploitation of allelopathy for weed control in annual and perennial cropping systems. J Chem Ecol 9:1001–1010

Putnam AR, Tang CS (1986) The science of allelopathy. Wiley, New York, NY

Qasem JR, Hill TA (1989) On difficulties with allelopathy methodology. Weed Res 29:345–347

Reader RJ (1991) Control of seedling emergence by ground cover: a potential mechanism involving seed predation. Can J Bot 69:2084–2087

Reigosa MJ, Pedrol N (2002) Allelopathy: from molecules to ecosystems. Science Publisher Inc, Enfield, NH

Rice EL (1983) Pest control with nature's chemicals: allelochemics and pheromones in gardening and agriculture. University of Oklahoma Press, Norman, OK

Rice EL (1984) Allelopathy. Academic Press, Orlando, FL

Rice EL (1995) Biological control of weeds and plant diseases. Advances in applied allelopathy. University of Oklahoma Press, Norman, OK

Rizvi SJH, Rizvi V (1992) Allelopathy: basic and applied aspects. Chapman & Hall, London

Rovira AD (1991) Rhizosphere research-85 years of progress and frustration. In: Keister DL, Cregan PB (eds) Beltsville symposium in agricultural research. 14. The rhizosphere and plant growth. Kluwer Academic Press, Dordrecht, pp 3–13

SAS Institute Inc (1999) SAS/STAT user's guide, version 8. SAS Publishing, Cary

Schmidt SK (1988) Degradation of juglone by soil bacteria. J Chem Ecol 14:1561–1571

Schmidt SK, Ley RE (1999) Microbial competition and soil structure limit the expression of allelopathy. In: Interjit, Dakshini KM, Foy CL (eds) Principles and practices in plant ecology: allelochemical interactions. CRC Press, Boca Raton, FL, pp 339–351

Shafer SR, Blum U (1991) Influence of phenolic acids on microbial populations in the rhizosphere of cucumber. J Chem Ecol 17:369–389

Shann JR, Blum U (1987) The uptake of ferulic and *p*-hydroxybenzoic acids by *Cucumis sativus*. Phytochemistry 26:2959–2964

Shilling DG, Liebl RA, Worsham AD (1985) Rye (*Secale cereale* L.) and wheat (*Triticum aestivum* L.) mulch: the suppression of certain broadleaved weeds and the isolation and identification of phytotoxins. In: Thompson AC (ed) The chemistry of allelopathy: biochemical interactions among plants. ACS symposium series, vol 268. American Chemical Society, Washington, DC, pp 243–271

Shilling DG, Worsham AD, Danehower DA (1986) Influence of mulch, tillage, and diphenamid on weed control, yield, and quality in no-till flue-cured tobacco (*Nicotiana tabacum*). Weed Sci 34:738–744

Siqueira JO, Nair MG, Hammerschmidt R, Safir GR (1991) Significance of phenolic compounds in plant-soil-microbial systems. Crit Rev Plant Sci 10:63–121

Sparling GP, Ord BG, Vaughan D (1981) Changes in microbial biomass and activity in soils amended with phenolic acids. Soil Biol Biochem 13:455–460

Sparling GP, Vaughan D (1981) Soil phenolic acids and microbes in relation to plant growth. J Sci Food Agric 32:625–626

Staman K, Blum U, Louws F, Robertson D (2001) Can simultaneous inhibition of seedling growth and stimulation of rhizosphere bacterial populations provide evidence for phytotoxin transfer from plant residues in the bulk soil to the rhizosphere of sensitive species? J Chem Ecol 27: 807–829

Steuter AA (1981) Water potential of aqueous polyethylene glycol. Plant Physiol 67:64–67

Stevenson FJ (1982) Humus chemistry. Wiley, New York, NY

Stowe LG, Osborn A (1980) The influence of nitrogen and phosphorus levels on the phytotoxicity of phenolic compounds. Can J Bot 58:1149–1153

Strobel BW (2001) Influence of vegetation on low-molecular-weight carboxylic acids in soil solution-a review. Geoderma 99:169–198

Stubbs TL, Kennedy AC, Reisenauer PE, Burns JW (2009) Chemical composition of residue from cereal crops and cultivars in dryland ecosystems. Agron J 101:538–545

Sylvia DM, Fuhrmann JJ, Hartel PG, Zuberer DA (2004) Principles and application of soil microbiology, 2nd edn. Prentice Hall, New Jersey, NJ

Tang CS, Waiss AC (1978) Short-chain fatty acids as growth inhibitors in decomposing wheat straw. J Chem Ecol 4:225–232

Tapin S, Sigoillot J-C, Asther M, Petit-Conil M (2006) Feruloyl esterase utilization for simultaneous processing of nonwood plants into phenolic compounds and pulp fibers. J Agric Food Chem 54:3697–3703

Taussky HH, Shorr E (1953) A microcolorimetric method for the determination of inorganic phosphorus. J Biol Chem 202:675–685

Teasdale JR (1988) Weed suppression by hairy vetch residues. Proc Northeast Weed Sci Soc 42:73

Thelander O (1985) Review of straw carbohydrate research. In: Hill RH, Munck L (eds) New approaches to research on cereal carbohydrates. Prog Biotech, vol I. Elsevier, New York, NY, pp 217–230

Turner JA, Rice EL (1975) Microbial decomposition of ferulic acid in soil. J Chem Ecol 1: 41–58

Vaughan D, Sparling GP, Ord BG (1983) Amelioration of the phytotoxicity of phenolic acids by some soil microbes. Soil Biol Biochem 15:613–614

Vermerris W, Nicholson R (2006) Phenolic compound biochemistry. Springer, Dordrecht

Walker TS, Bais HP, Grotewold E, Vivanco JM (2003) Root exudation and rhizosphere biology. Plant Physiol 132:44–51

Waller GR (1987) Allelochemicals: role in agriculture and forestry. ACS symposium series, vol 330. American Chemical Society, Washington, DC

Whitehead DC, Dibb H, Hartley RD (1981) Extractant pH and the release of phenolic compounds from soils, plant roots and leaf litter. Soil Biol Biochem 13:343–348

Wickliffe WB II (1999) Enhancing weed suppression by increasing rye cover crop seedling rates in no-till tobacco production. North Carolina State University Thesis, Raleigh, NC

Willard JI, Penner D (1976) Benzoxazinones: cyclic hydroxamic acids found in plants. Residue Rev 64:67–76

Williamson GB, Weidenhamer JD (1990) Bacterial degradation of juglone. Evidence against allelopathy? J Chem Ecol 16:1739–1752

Wilson RE, Rice EL (1968) Allelopathy as expressed by *Helianthus annuus* and its role in old-field succession. Bull Torrey Bot Club 95:432–448

Worsham AD (1989) Current and potential techniques using allelopathy as an aid in weed management. In: Chou CH, Waller GR (eds) Phytochemical ecology: allelochemicals, mycotoxins, and insect pheromones and allomones. Academia Sinica Monograph Series No. 9. Institute of Botany, Taipei, ROC, pp 275–291

Worsham AD (1990) Weed management strategies for conservation tillage in the 1990's. In: Mueller JP, Wagger MG (eds) Conservation tillage for agriculture in the 1990's: proceedings of the 1990 southern region conservation tillage conference, pp 42–47. Special Bulletin 90-1. North Carolina State University, Raleigh

Wu H, Haig T, Pratley J, Lemerle D, An M (2000a) Allelochemicals in wheat (*Triticum aestivum* L.): variation of phenolic acids in root tissue. J Agric Food Chem 48:5321–5325

Wu H, Haig T, Pratley J, Lemerle D, An M (2000b) Distribution and exudation of allelochemicals in wheat *Triticum aestivum*. J Chem Ecol 26:2141–2154

Wu H, Haig T, Pratley J, Lemerie D, An M (2001a) Allelochemicals in wheat (*Triticum aestivum* L.): cultivar differences in exudation of phenolic acids. J Agric Food Chem 49: 3742–3745

Wu H, Haig T, Pratley J, Lemerle D, An M (2001b) Allelochemicals in wheat (*Triticum aestivum* L.): variation of phenolic acids in shoot tissue. J Chem Ecol 27:125–135

Wu H, Haig T, Pratley J, Lemerle D, An M (2001c) Allelochemicals in wheat (*Triticum aestivum* L.): production and exudation of 2,4-dihydroxy-7-methoxy-1,4-benzoxazin-3-one. J Chem Ecol 27:1691–1700

Xu JM, Tang C, Chen ZL (2006) The role of plant residues in pH change of acid soils differing in initial pH. Soil Biol Biochem 38:709–719

Yenish JP, Worsham AD, Chilton WS (1995) Disappearance of DIBOA-glucoside, DIBO, and BOA from rye (*Secale cereale* L.) cover crop residue. Weed Sci 43:18–20

Yongoing MA (2005) Allelopathic studies of common wheat (*Triticum aestivum* L). Weed Biol Manag 5:93–104

Zeng RS, Mallik AU, Luo SM (2008) Allelopathy in sustainable agriculture and forestry. Springer, New York, NY

Chapter 4
Phase III: Summing Up

Abstract This chapter provides an extended listing of if-then hypotheses for each objective of this research as a way of summarizing observations and/or conclusion from both laboratory and field bioassays. Provides both cons and pros for the following central questions: (a) How likely are the necessary phenolic acid concentrations and environmental conditions present in wheat no-till crop systems for the inhibition of weed seedling emergence to occur? and (b) Do phenolic acids have a dominant role in regulating broadleaf weed seedling emergence or are phenolic acids just one component of a larger promoter/modifier/inhibitor complex that regulates weed seedling emergence in wheat no-till crop systems? Proposes a modified paradigm for the study of plant–plant allelopathic interactions, suggests several ways by with this modified paradigm may be implemented, and updates the criteria to establish plant–plant allelopathic interactions consistent with this modified paradigm.

4.1 Hypotheses

In the first chapter I stated that:

> The science of allelopathy, like all other sciences, has a prescribed methodology by which problems are to be addressed and solved, the "Scientific Method". Once a problem has been identified, this method requires that alternative hypotheses, tentative explanations, be generated which can then be experimentally tested. Occurrence (acceptance) or non-occurrence (rejection) of predictions deduced for each hypothesis is then determined by means of experiments. At that point new hypotheses are formulated or old ones are modified. Furthermore, science progresses not by trying to confirm hypotheses but by attempting to falsify them since it is usually possible to find at least some confirmatory evidence for any hypothesis, but one solid piece of negative data refutes a hypothesis completely. (Blum 2007; Used with permission of Science Publishers)

I also stated that in biology the most frequently used tools for experimentally testing hypotheses are field and/or laboratory bioassays. Having said that, where are the hypotheses in this book?

First, however, what is a hypothesis? A hypothesis is a tentative statement of a potential explanation of some phenomenon, processes, or event that is testable

U. Blum, *Plant–Plant Allelopathic Interactions*, DOI 10.1007/978-94-007-0683-5_4,

and that frequently includes predictions as to outcome if the hypothesis is true, i.e., essentially a conditional statement. For example, if this is true, then this is also true or this will happen, the "if" part of a conditional statement is the hypothesis and the "then" part is the conclusion. Since conditional statements may be true or may be false, we will not know which conditional statement is true and which is false until adequately tested. Finally, hypotheses to be tested are written or modified before the actual experiments are carried out, not after.

This book describes the results/conclusions of a number of laboratory and field bioassays (i.e., experiments) each with its own set of underlying hypotheses. Because this book is a retrospective analysis it is impossible to list all the hypotheses that have been directly or indirectly tested for each reference cited. In fact, even if I wanted to it would not be possible since specific hypotheses are frequently not even stated in the plant–plant allelopathic literature. Let me suggest some reasons for that:

a. the research presented is not experimental but observational or descriptive,
b. the researcher assumes them to be "intuitively obvious" to the reader (i.e., the reader is assumed to be sufficiently knowledgeable to determine what they are),
c. when researchers design experiments the underlying hypotheses are inherent in the design and just not expressed,
d. limited space in journals does not allow for lengthy discourse on underlying assumptions/hypotheses of experimental design, etc., and/or
e. if a researcher does not believe in the "Scientific Method" per se as stated above (i.e., the "Scientific Method" is much too simplistic) and does not have an overall defined research strategy then each research project has its own strategy and the correct strategy may not be known until the work is done. For additional details see Laidler (1998).

What I will do here is to provide a few central hypotheses for each objective in this book, minimizing repetition as much as possible. However, the reader should recognize that these hypotheses were not formulated prior to actual experimentation since experimental results reported here have already been published. Furthermore, the reader will find that a number of these hypotheses are intuitively obvious, biased towards what I think are the central issues of this research, and represent provisional conclusions for this research, which may be true or may be false. Additional research will be required to ultimately determine which of these are true (accepted) and which of these are false (rejected). We may tentatively assume them to be true until proven false. In spite of this, restating the objectives as hypotheses is constructive in that it provides an additional way of focusing on the central issues of the research described in this book.

4.1.1 Plant–Plant Allelopathic Interaction. Phase I: The Laboratory

4.1.1.1 To Determine the Effects and the Duration of Effects of Phenolic Acid on Seedlings in Nutrient Culture (Section 2.4.1)

a. If effects of simple phenolic acids (e.g., ferulic, *p*-coumaric, vanillic, and *p*-hydroxybenzoic acids) on cucumber seedling (e.g., reduction in net nutrient and water uptake, water utilization, photosynthesis, and growth) are typical of phenolic acid effects observed for other sensitive seedling taxa, then cucumber seedlings will be a good model seedling for determining how simple phenolic acids may potentially influence seedling processes and modify seedling behavior.
b. If simple phenolic acids in Hoagland's nutrient solution reduce net water and nutrient uptake of phenolic acid-sensitive seedlings and this behavior is directly related to concentration and inversely related to pH of the nutrient solution, then increasing concentrations of phenolic acids in a solution with a given pH will increase the observed reduction in net nutrient (e.g., ammonium, nitrate, potassium, and phosphorous) and net water uptake, and eventually, given enough time, reduce water utilization, transpiration, photosynthesis, carbon allocation to the roots, leaf expansion, root growth, biomass accumulation, etc. Furthermore, for a given concentration of phenolic acid as the pH of the nutrient solution is increased (made less acidic) or decreased (made more acidic) observed effects will decline or increase, respectively.
c. If root contact with simple phenolic acids is more important than root uptake of simple phenolic acids, then the relationships between phytotoxicity (i.e., effects) and root contact will be more consistently related (i.e., significant with higher r or r^2 values) than the relationships between phytotoxicity and root uptake.
d. If lasting effects of simple phenolic acids on seedling processes requires continuous root contact, then affected processes will recover as root contact with phenolic acids is reduced or eliminated. Processes will completely recover given sufficient time, however, anatomical or morphological modifications, once established, will not be reversed.
e. If effects of simple phenolic acids are local, not systemic, then effects of phenolic acids will increase or decrease as the proportion of a root system in contact with phenolic acids is increased or decreased, respectively.
f. If sensitive seedlings adjust and acclimatize their behavior to environmental changes and/or stresses, then environmental changes and/or stresses will modify seedling responses to simple phenolic acids.
g. If simple phenolic acids (e.g., ferulic, *p*-coumaric, vanillic, and *p*-hydroxybenzoic acids) interact with the same primary sites of sensitive seedling root membranes, then the effects of individual phenolic acids in a mixtures of simple phenolic acids will be additive at very low concentrations and increasingly more antagonistic with increasing concentrations as the primary sites becomes saturated and

phenolic acids compete for sites. Their effects will never be synergistic unless the site and/or mode of action are modified. Furthermore, effects of mixtures of phenolic acids will still be evident even when individual phenolic acid concentrations within the mixture are well below their effective concentrations, as long as, the total concentration of the mixture is at an effective (e.g., inhibitory) level. Or stated another way: If the effects of individual simple phenolic acids are additive or partially antagonistic, then increasing the number of simple phenolic acids in a nutrient culture reduces the concentration of each phenolic acid required for a given percent inhibition.

h. If primary sites on root membranes are selective for different types and states (e.g., benzoic acids vs. cinnamic acids, phenolic acids with different side groups, or protonated vs. ionized) of simple phenolic acids and membrane perturbations vary with site, phenolic acid, and state of phenolic acid, then effects of individual phenolic acids will vary accordingly. Furthermore in mixtures of phenolic acids the contribution of an individual phenolic acid to the total effect of the mixture will depend on site preference for and state of that phenolic acid and the nature of its perturbation.
i. If simple phenolic acids are sorbed to cell walls and/or taken up by root cells and subsequently utilized in lignin synthesis (i.e. secondary cell walls), then cell expansion (i.e., root growth) will be inhibited, and
j. If seedling roots can modify their immediate environment by changing nutrient solution pH and thus change the concentrations of active simple phenolic acids in the nutrient solution and if roots can neutralize phenolic acids through sorption to cell walls, incorporation into lignin, glucosylation, metabolism, and/or storage in vacuoles, then seedling roots can directly influence the magnitude of primary and secondary effects of simple phenolic acids through feedback regulation.

4.1.1.2 To Determine the Effects of Seedlings, Mixtures of Phenolic Acids, and Microbes on Phenolic Acid Concentrations in Nutrient Culture (Section 2.4.2)

a. If simple phenolic acids can be taken up by roots, then the rate of depletion will depend on the type of uptake (i.e., passive or active), the root environment (e.g., phenolic acid concentration, aeration, and pH), root type, root surface area, and acclimation and "health" of seedling roots and shoots.
b. If simple phenolic acids can be utilized as a carbon source by microbes located within the nutrient solution and/or associated with the rhizoplane (root surface), then microbial populations that can utilize phenolic acids as a carbon source within the nutrient solution and on the rhizoplane will increase and the rate of phenolic acid depletion will increase over time. The actual rate of depletion will depend on the type, rate of induction/selection, and reproduction of microbes, number of initial microbes present, the root and microbe environments, as well as root uptake. Phenolic acid breakdown products will accumulate in the nutrient solution, particularly when nutrient solutions are low in oxygen.

c. If seedling roots take up simple phenolic acids and uptake is related to rates of metabolic activity (i.e., active uptake), then depletion of phenolic acids from nutrient solution will be modified by the presence or absence of light, changing temperatures, and levels of aeration, but will be independent of concentration as long as concentrations are not limiting and feedback inhibition is absent. Given the above environmental modifiers depletion by roots will vary with time of day.
d. If uptake of simple phenolic acid by roots is passive (not related to metabolic activity), then depletion from nutrient solution will be concentration dependent as long as external cell concentrations are higher than internal cell concentrations and will not be modified by the presence or absence of light, changing temperatures, and levels of aeration. Depletion from nutrient solution will be concentration dependent no matter the time of day as long as the proper gradient exists.
e. If root contact and uptake (symplastic and apoplastic) of simple phenolic acids by sensitive seedling occurs primarily when the carboxy sites of phenolic acids are protonated (i.e., neutral) and the simple phenolic acids in question have a pK_a of approximately 4.5, then root contact and uptake will primarily occur under acidic conditions. Or stated another way: If root uptake of phenolic acids is inversely related to pH, then depletion of phenolic acids from nutrient solutions will increase with decreasing (more acidic) pH.
f. If roots (e.g., cell walls, membrane sites) of sensitive seedlings adjust and/or acclimate to the frequency of simple phenolic acid treatments, then rates of depletion of phenolic acids from nutrient solution will be modified by repeated phenolic acid treatments.
g. If membrane sites for uptake are selective for type and state of simple phenolic acids and uptake occurs at the same sites on root membranes, then depletion of individual phenolic acids from a mixture of phenolic acids in nutrient solution, as long as total concentration is well below saturation of the sites, will have rates consistent with their type and state and will be independent of other phenolic acids present (i.e., additive uptake). As the total concentration increases and begins to saturate uptake sites depletion of phenolic acids from nutrient solution will change (e.g., differential antagonistic uptake) since the rate of change of a given phenolic acid will depend on how selective the uptake site is for a particular phenolic acid, and
h. If simple phenolic acids are sorbed to cell walls, incorporated into lignin, glucosylated, metabolized, and sequestered into vacuoles of root cells, then a large fraction of phenolic acids interacting with roots will be fixed/retained in the root tissue.

4.1.1.3 To Determine the Interactions of Phenolic Acids with Sterile and Non-sterile Soils (Section 2.4.3)

The focus here is more on the development of tools to recover available fractions of phenolic acids than it is on identifying and characterizing patterns and mechanisms of phenolic acid sorption in soils.

a. If simple phenolic acids added to soils are sorbed or bound onto or into soil particles (e.g., clays, and organic particles) by a range of mechanisms (e.g., H bonding, multivalent cation bridging, ligand exchange, or polymerization) with different bonding types and/or strengths, then different extraction procedures utilizing a range of inorganic and organic solvents of various strengths, environments (e.g., pH, temperatures, and pressures), and extraction times will recover different fractions or combinations of fractions of phenolic acids from soil based on bonding types and/or strengths. Note: For simplicity we have classified the soil fractions in three groups (i.e., "free" (soil solution), reversibly sorbed, and irreversibly sorbed).
b. If microbes utilize certain fraction or fractions of the simple phenolic acids in soil at different rates and over different time intervals (e.g., utilization of "free" fraction > reversibly sorbed fraction > irreversibly sorbed fraction) and given that hypothesis a. is true, then extraction procedures can be found to recover and isolate the specific fraction or fractions utilized by soil microbes.
c. If hypothesis a. is true and neutral EDTA (ethylenediaminetetraacetic acid) and NaOH extractions recover the microbial available fractions, and the microbial available fractions plus a portion of the short-term non-available (irreversible or fixed) fraction, respectively, then the following will also be true. As soil microbes utilize (deplete) available simple phenolic acids ("free" and reversibly sorbed) from soil the amount of the phenolic acids recovered by both neutral EDTA and NaOH extractions will decline proportionally. Once soil available phenolic acids have been depleted by microbes recovery by neutral EDTA extractions will be zero but recovery of phenolic acids by NaOH extractions will continue thereafter at a constant level (i.e., recovery of irreversible bound forms). Note: EDTA and NaOH extractions at each time interval were made on different subsoil samples.
d. If hypothesis a. is true and the fractions of phenolic acid utilized by microbes are identical or very similar to what interacts with or are taken up by roots, then the same extraction procedures can be used to characterize/quantify available phenolic acids in soil for both microbes and roots.
e. If water extractions and ambient soil solutions from identical soil samples contain approximately equal concentrations of simple phenolic acid, then water extractions will provide an estimate of the "free" phenolic acids in soil solution. If water extractions contain higher concentrations of simple phenolic acids than ambient soil solutions, then water extractions will recover the "free" fraction and some of the more easily desorbed (reversibly sorbed) simple phenolic acids.
f. If water extractions recover "free" phenolic acids in soil solution and if neutral EDTA extractions recover all microbe and root available phenolic acids, then the difference between water and neutral EDTA extractions will represent the amount of reversibly sorbed phenolic acids in soil.
g. If reversible sorption of simple phenolic acids to soils, microbes, and roots is partly due to cationic bridging between negatively charged surfaces and negatively charged phenolic acids, then increasing or decreasing multivalent cations in the soil will increase or decrease reversible sorption, respectively.

h. If reversible sorbed phenolic acids decline as "free" phenolic acids are depleted from the soil solution, then reversibly sorbed phenolic acids will be readily available to interact with microbes and roots.
i. If reversibly sorbed phenolic acids decline as phenolic acid concentrations decline in the soil solution, then reversibly sorbed phenolic acids will also increase as the phenolic acid concentrations of the soil solution increase, and
j. If after phenolic acid enrichment but before microbial utilization it is observed that the recovery of simple phenolic acids from sterilized and non-sterilized soil are approximately equal, then the sterilization procedure used will be a good candidate for characterizing/quantifying "free", reversible sorbed, and irreversible sorbed phenolic acids in soil in the absence of microbes (i.e., in sterilized soil).

4.1.1.4 To Determine the Effects of Phenolic Acids on Bulk-Soil and Rhizosphere Microbial Populations (Section 2.4.4)

a. If bulk-soil and rhizosphere/rhizoplane microbes can utilize simple phenolic acids as a carbon source, then simple phenolic acid enrichment, depending on concentration and soil physicochemical environment, will stimulate induction/selection of bulk-soil and rhizosphere/rhizoplane microbes that can utilize phenolic acids as a carbon source.
b. If induction/selection of phenolic acid-utilizing microbes is taxon specific, then the rates and magnitude of selection/induction will depend on the type of microorganism (e.g., bacteria, fungi, and actinomycetes) present, as well as, the soil physicochemical environment (e.g., pH, nutrition, moisture, and temperature).
c. If enriching soils with simple phenolic acids leads to a strong induction/selection pressure for phenolic acid-utilizing microbes, then this induction/selection will not only lead to a dramatic increase in populations (taxa) that can utilize phenolic acids but also to a decline in species diversity of the microbial community in the soil.
d. If microbes vary in their preferences for carbon sources, then the presence of other more preferred carbon sources, other than simple phenolic acids, will reduce the rate of microbial utilization of simple phenolic acids as a carbon source. Note: Preferences can also occur between different simple phenolic acids, and
e. If soil and rhizosphere/rhizoplane microbes that have been induced/selected by a specific phenolic acid can still utilize other phenolic acids, then these microbes function as generalists for phenolic acids. If they cannot utilize other phenolic acids, then they function as specialists for that phenolic acid.

4.1.1.5 To Determine the Effects and the Duration of Effects of Phenolic Acids on Seedlings in Soil Culture (Section 2.4.5)

a. If secondary effects (e.g., reduction in water utilization and leaf area expansion) of simple phenolic acids are closely or directly linked to growth and development of seedlings, then any variation in the growth environment (e.g., soil type, nutrition, moisture, temperature, and pH) will modify the rate (i.e., time) and magnitude of inhibition and/or subsequent recovery of seedling.

b. If simple phenolic acids, in addition to their different potential phytotoxicities, are sorbed and bound in/on soil particles, leached, utilized by microbes, moved vertically and horizontally by capillary action or mass flow, and taken up by roots at different rates and these rates are determined by soil biotic and physico-chemical properties, then the rate (i.e., time) and magnitude of inhibition and/or subsequent recovery of affected seedlings will likely be different (e.g., ranging from slightly to substantially different) for every soil sample tested (i.e., there is a high probability that every soil sample from/in nature is unique, if not completely at least to some extent) and for each individual phenolic acid tested.
c. If soils contain a large enough number of available active simple phenolic acids, each at concentrations well below their individual inhibitory levels, then the effects of this complex (mixture) of non-inhibitory phenolic acids will be inhibitory, and
d. If soils contain a range of stimulatory, neutral, and inhibitory concentrations of organic and inorganic compounds, in addition to active simple phenolic acids, and the action of the various individual components in this promoter/modifier/inhibitor complex are synergistic, neutral, additive, or antagonistic, then inhibitory effects of simple phenolic acids will only be observed/identified/characterized if phenolic acids dominate the complex or are a proxy for the inhibitor portion of the complex.

4.1.1.6 To Determine Relationships Between Phenolic Acids-Utilizing Microbes and Phenolic Acid Inhibition (Section 2.4.6)

a. If available phenolic acid concentrations are high enough or movement to roots fast enough so that the effects of simple phenolic acids on seedling and rhizosphere/rhizoplane microbes are independent, then simultaneous stimulation of phenolic acid-utilizing microbes in the rhizosphere/rhizoplane and inhibition of seedlings will occur.
b. If conditions for a. are not true and if concentrations of available phenolic acids in soil are closely or directly linked to inhibition and recovery of sensitive seedlings (e.g., leaf expansion), then increases or decreases in phenolic acid-utilizing bulk-soil and rhizosphere/rhizoplane microbes will decrease or increase seedling inhibition and increase and decrease recovery, respectively.
c. If the initial breakdown products of phenolic acids are other phenolic acids, then the effective inhibitory dose of phenolic acids on sensitive seedlings will not only be determined by the rates of microbial utilization/synthesis of phenolic acids but also by the rates of synthesis/utilization of phenolic acid breakdown products, and

d. If phenolic acid-utilizing microbes utilize carbon sources preferentially, then the addition (presence) of more readily utilizable carbon sources other than simple phenolic acids would reduce phenolic acid utilization and increase seedling inhibition for a given phenolic acid concentration. Note: Preferences can also occur between different simple phenolic acids with similar or different phytotoxicities.

4.1.1.7 To Determine the Effects of Seedling-Microbe-Soil Systems on the Available Concentrations of Phenolic Acids in Soil Solutions (Section 2.4.7)

a. If the available simple phenolic acid concentrations in field soil are fairly constant over a growing season but simple phenolic acids in laboratory seedling-microbe-soil systems decline rapidly after each addition of phenolic acids, then multiple additions or continuous additions of phenolic acids to seedling-microbe-soil systems will more closely mimic field soil systems.
b. If environmental conditions (e.g., moisture, nutrition, pH, and temperature) are above the minimum required (adequate or better than adequate) but below inhibitory levels for seedling and microbial growth and reproduction, then enrichment of soil with simple phenolic acids will lead to a rapid decline of available phenolic acids in the soil solution, the rate being determined by the microbial taxa present, root surface area in contact with phenolic acids, soil type, and environmental conditions. The decline will primarily be result of microbial utilization, but also a result of root uptake and soil sorption. Note: Under adequate environmental conditions phenolic acid enrichment will lead to a rapid selection/induction of active phenolic acid-utilizing microbes.
c. If simple phenolic acids inhibit water utilization (i.e., transpiration) of sensitive seedlings, if inhibition of water utilization by phenolic acids is concentration dependent, and if soil solution concentrations of available phenolic acids are inversely related to soil moisture, then inhibition of water utilization and phenolic acid concentrations in the soil solution will also be inversely related. In other words, the higher the inhibition of water utilization and transpiration the lower the concentration of available phenolic acids in the soil solution, and
d. If available simple phenolic acid concentrations vary with soil moisture, then μM or mM will be biologically more meaningful than μmol/g soil.

4.1.1.8 To Compare the Effects of Phenolic Acids on Seedlings in Nutrient and Soil Culture (Section 2.4.8)

If seedling-microbe-soil systems have different or stronger sinks (e.g., microbial utilization, root uptake, and soil sorption), slower root growth, and reduced root contact with simple phenolic acids (e.g., resistance of movement for simple phenolic acids through soil capillaries, and slower root growth) than occur in nutrient culture systems then:

a. initial development and maximum seedling inhibition by simple phenolic acids, and subsequent recovery after phenolic acid depletion in nutrient culture systems will be much more rapid in nutrient culture than in soil culture,
b. for a given percent seedling inhibition lower concentrations of simple phenolic acids will be required in nutrient culture than in soil culture, and
c. seedling inhibition observed in nutrient culture will more closely represent the maximum potential phytotoxicity of simple phenolic acids while inhibition in soil systems represents a soil modified phytotoxicity. Note: A soil modified phytotoxicity should be more representative of field soils.

4.1.1.9 To Determine the Effects of Phenolic Acids at Various Life Stages (Section 2.4.9)

If the weakest links of a plant's life cycle to a range of environmental stressors are early seedling establishment (e.g., particularly the transition from heterotrophic to the autotrophic stage), and reproduction (e.g., flower initiation, pollination, and seed and fruit development), then the greatest effects of phenolic acids, like all other stressors, will occur at these life cycle stages.

4.1.2 Plant–Plant Allelopathic Interactions Phase II: Field/Laboratory Experiments

For the hypotheses of Phase I (Section 4.1.1; Chapter 2) I was able to focus directly on the roles/functions of phenolic acids because phenolic acid treatments and the environments of our model systems were directly manipulated in our experiments. For Phase II (field/laboratory experiments; Chapter 3) I could not single out phenolic acids as a treatment variable since the tissue/residues and their subsequent organic fragments and leachates were composed of phenolic acids imbedded in unknown mixtures of other organic and inorganic compounds. Furthermore, we had very little control over the physicochemical environment of the experimental plots. For the associated laboratory experiments we could manipulate the treatments and the environments, but even for these experiments we were testing unknown mixtures of organic and inorganic compounds, not just phenolic acids. Thus a number of the following hypotheses are much more indirect when it comes to the roles and functions of phenolic acids.

4.1.2.1 To Characterize Phenolic Acid in Soils of No-Till and Conventional-Till Systems and to Establish Correlations Between Easily Obtained Soil Characteristics and Phenolic Acids in Soils (Section 3.4.1)

a. If phenolic acids are the primary inhibitory component in soils, and soils contain a large number of active phenolic acids and phenolic acid moieties, and the inhibitory action of individual available phenolic acids are additive or partially antagonistic depending on concentration, pH, etc., then estimates of concentrations for all active individual and active total phenolic acids will be required to understand the inhibitory behavior of phenolic acids on sensitive seedlings. Note: Since presently we are unable to determine active fractions of phenolic acids in soils the best that can be achieved is to determine the available (available/active) fractions.

b. If neutral EDTA extractions (our standard; for justification see Section 2.4.3.1) and water-autoclave extractions both consistently recover roughly equivalent quantities of available ("free" and reversibly sorbed) individual phenolic acids from soil, then water-autoclave extractions can also be used to obtain an estimate of available total phenolic acid content of soil. Note: Since neutral EDTA reduces the Folin & Ciocalteu's phenol reagent, EDTA extracts cannot be used to estimate available total phenolic acid content.
c. If the Folin & Ciocalteu's phenol reagent primarily interacts with phenolic acids and phenolic acid moieties in water-autoclave soil extracts and the resulting quantitative total phenolic acid values of these extracts are consistently related to individual or sums of common individual phenolic acids in those extracts, then the phenol reagent will provide reliable values for total available phenolic acids in soil extracts. Note: These values will be relative not absolute, because these values are based on the color development of one phenolic acid (phenolic acid equivalence) and different phenolic acids produce different shades of blue with the Folin & Ciocalteu's phenol reagent.
d. If the Folin & Ciocalteu's phenol reagent interacts not only with phenolic acids but also with various other organic and inorganic compounds, then comparisons of total phenolic acid for extracts from different taxa of plant tissues/residues (e.g., small grains vs. clovers) and soils containing different taxa of plant tissues/residues will not be directly comparable since their composition of organic and inorganic compounds will be different.
e. If available phenolic acids in soil come from shoot and root tissues/residues and their release from tissues/residues is highly variable, and if phenolic acids once released to the soil are differentially sorbed to soil particles, utilized by soil microbes, leached, and taken up by living roots, then the importance/concentration of available individual phenolic acids or total phenolic acid in shoot and root tissues/residues will provide little, if any, insight regarding the importance/concentration of actual available phenolic acids in soil.
f. If available phenolic acids come from shoot and root tissues/residues and recalcitrant organic matter, then the actual observed distribution pattern of phenolic acids in the soil profile will be an integration of the patterns generated by these sources. See 1–3 below for potential distribution patterns of individual sources. Note: The greatest change in distribution patterns of phenolic acids (and other organic and inorganic compounds) due to shoot and root tissue/residues will occur shortly after glyphosate desiccation, as long as adequate rain fall events occur and adequate soil moisture exists. Once the readily available phenolic acids

(and other organic and inorganic compounds) have been released from shoot and root tissues/residues soil phenolic acid concentrations (and other organic and inorganic compounds) should remain fairly constant or change/fluctuate slowly over time.

1. If available phenolic acids in soil come from leachates of shoot tissues/residues above and on the surface of the soil then:

    a. concentrations of available phenolic acids will be highest near the surface and/or the upper region of the soil and decline rapidly with soil depth,
    b. the actual horizontal and vertical distribution of available phenolic acids in the soil will be determined by the distribution of shoot tissues/residues above and on the soil surface, the timing, frequency, and magnitude of rainfall events, and the action of soil processes, and
    c. the highest observable concentrations of available phenolic acids at and just below the soil surface will occur shortly after glyphosate desiccation of the cover crop and the first adequate rainfall event. Note: a, b, and c are also true for all other organic and inorganic compounds.

2. If available phenolic acids in soil come from root tissues/residues, then the distribution of available phenolic acids will be consistent with root tissue/residue distribution in the soil, movement of gravitational and capillary water, mass flow of soil solutions driven by "transpirational pull", and the action of soil processes. Concentrations released will be highest shortly after glyphosate desiccation. Note: This is also true for all other organic and inorganic compounds, and
3. If available phenolic acids in soil come from recalcitrant organic matter, then soil concentrations of available phenolic acids will be determined by organic matter distribution, faunal and microbial activity, movement of gravitational and capillary water, mass flow of soil solutions driven by "transpirational pull", and the action of soil processes. Furthermore the local concentrations will tend to be fairly constant over the growing season as long as the soil physicochemical environment (e.g., soil moisture, nutrition, temperature, and pH) is sufficient for the release of phenolic acids by faunal and microbial activity. Note: This assumes that more phenolic acids are released than are utilized by these organisms. The above statement is also true for all other organic and inorganic compounds, and

g. If cover crop shoot and root tissues/residues are tilled into the soil, then the concentration of available phenolic acids and other organic and inorganic compounds that have accumulated at or near the surface of the soil will be redistributed/equalized/diluted throughout the tilled soil profile and any inhibition zone at/near the surface of the soil will be destroyed. Furthermore, the future source, i.e., shoot and root tissues/residues, of phenolic acids and other organic and inorganic compounds will also be redistributed/equalized/diluted throughout the soil profile.

4.1.2.2 To Determine If Soil Extracts Could Be Used Directly in Laboratory Bioassays for the Detection of Allelopathic Activity (Section 3.4.2)

a. If germination and early seedling growth are determined by a variety of chemical and physical soil characteristics and these act or interact in various ways (i.e., their actions are independent, synergistic, additive, or antagonistic), then the role of allelopathic agents, such as phenolic acids, cannot be viewed in isolation since the resulting effects are due to the action and interactions of the physical environment and other organic and inorganic compounds present. More specifically, phenolic acids are just one component among many other active organic and inorganic compounds in soil extracts (i.e., soil extracts being proxies for soil solutions).
b. If simulated (model) soil extracts with the appropriate combinations and concentrations of phenolic acids, pH, solute potential, etc. are created in a manner which would allow these factors (i.e., phenolic acids, pH, solute potential, etc.) to be independently manipulated, then it will be possible to identify the potential contribution of each factor to the resulting observed effects and identify possible co-linearities of factors.
c. If dose responses (i.e., serial dilutions tested on sensitive species of interest) for individual soil extracts generate significant slopes that are representative of the stimulatory or inhibitory potential of the extracts tested then:

    1. each significant slope is a proxy for biological activity of a soil extract,
    2. significant slopes for a range of extracts representing a field plot can be regressed against soil extract characteristics (e.g., total phenolic acid, extract pH, solute potential, and nutrient levels) of the undiluted extracts to identify which extract characteristics contributed to the effects observed, and
    3. the percent of significant extract slopes obtained for a field plot will provide an estimate of the potential allelopathic activity within that field plot. Note: This assumes that the soil samples extracted were representative of the field plot under study, and

d. If, for example, the significant slopes for soil extracts are significantly related to total available phenolic acids of the undiluted soil extracts, then phenolic acids are the dominant inhibitors or total phenolic acid is a proxy for all the inhibitors in the extracts.

4.1.2.3 To Characterize How Cover Crop Residues in No-Till Systems Affect Early Emergence of Broadleaf Weeds and to Establish and Characterize Potential Relationships Between Early Broadleaf Weed Seedling Emergence and the Physical and Chemical Environments Resulting from the Presence of Cover Crop Residues (Section 3.4.3)

a. If cover crop residues modify the physicochemical environment (e.g., temperature, soil moisture, light intensity at the soil surface, organic matter, nutrients, promoters, and inhibitors) in the immediate areas around seeds or seedling roots, then germination and weed seedling emergence, depending on species sensitivities to such physicochemical factors, will be modified accordingly.

b. If each organic and inorganic compound released from cover crop tissues/residues has the potential to be stimulatory, neutral, or inhibitory to seed or seedling processes depending on its concentration, physicochemical state, and environment, and the actions of individual compounds within a mixture of compounds has the potential to be synergistic, additive or antagonistic to seed or seedling processes, then identifying a single causative agent for an observed stimulatory or inhibitory effect is inappropriate unless that compound dominates over all other compounds in the mixture.
c. If the organic and inorganic compounds released from desiccated cover crops vary in amount and/or composition with species, age, "health", growth environment, management, and time of desiccation, then the effects of the resulting organic and inorganic compounds including phenolic acids on sensitive weed seedling emergence will vary accordingly.
d. If organic and inorganic compounds dominated by phytotoxins including phenolic acids are released from cover crops by leaching shortly after cover crops have been desiccated by glyphosate, then inhibition of seedling emergence of sensitive weed species will be most evident at that point in time assuming environmental conditions are appropriate for seedling emergence. Or stated another way: If cover crops are desiccated shortly before weed seedling emergence, then organic and inorganic compounds including phenolic acids released from cover crops by the first adequate rainfall event will have the greatest chance of influencing sensitive weed seedling emergence, and
e. If the release of organic and inorganic compounds dominated by phytotoxins including phenolic acids decline over time and the causative agents are not systemic in action, then weed seedling emergence will partially or completely recover, or even be enhanced over time, assuming environmental conditions are appropriate for seedling emergence.

4.1.2.4 To Characterize Cover Crops and Cover Crop Residues and How These May Potentially Modify the Soil Environment (Section 3.4.4)

a. If different types (i.e., taxa with different morphologies and chemical compositions) and amounts of glyphosate-desiccated cover crop residues are present on and in no-till soils, then factors such as solar radiation at the soil surface, soil temperature, soil moisture, soil pH, organic and inorganic compounds including phenolic acids in the bulk soil and soil solution, and/or organic particles and fragments in and on soil will vary.
b. If maximum potential release of organic and inorganic compounds occurs shortly after cover crop desiccation, then the most dramatic change in soil physicochemistry will occur right after the first adequate rainfall event following desiccation.

c. If organic and inorganic compounds including phenolic acids come from shoot residues, then concentrations of these compounds will be highest near the surface of the soil because of the compressed distribution of aboveground shoot residues at the soil surface. If organic and inorganic compounds come from root residues or recalcitrant organic matter, then the distribution will vary with their distributions. Note: Since in most instances shoots, roots, and recalcitrant organic matter are present the actual distribution will be some blend of the three (see Section 4.1.2.1 f. for additional details).
d. If after the initial release of organic and inorganic compounds from shoot and root residues there continues to be a slow release of compounds by way of decay processes (i.e., physical fragmentation, faunal and microbial activity, and leaching), then the bulk soil and soil solution concentrations will stabilize and input and losses of compounds will remain fairly constant as long as environmental conditions are appropriate and do not change dramatically, and
e. If living aboveground biomass is tilled into the soil, then the biomass and its associated compounds including phenolic acids will be diluted by the vertical and horizontal redistribution of the biomass throughout the soil. In addition, the tillage will bring to the surface weed seeds from the lower part of the soil profile (seed bank), increase aeration, reduce soil compaction, and create rough soil surfaces (i.e., desirable germination sites) that can potentially enhance germination and seedling emergence.

4.1.2.5 To Determine Under Controlled Conditions How Effects of Shoot Cover Crop Residues Taken from the Field Change with Time After Desiccation and How Such Effects Are Modified by Temperature, Moisture, and Nitrogen Levels (Section 3.4.5)

a. If in the field organic and inorganic compounds including phenolic acids decline over time in cover crop residues after glyphosate desiccation, and phenolic acids are the primary inhibitors or a proxy for all the inhibitors present in the residue, then inhibition of weed seedling emergence by residues taken at various times after glyphosate desiccation and tested under laboratory conditions will decline and be significantly related to phenolic acid content of the residue, assuming that environmental conditions for the laboratory bioassays are appropriate and remained constant over time.
b. If weed seedling emergence is modified by physicochemical factors such as temperature, moisture, pH, and organic and inorganic compounds and these factors act in a synergistic, additive, independent, or antagonistic manner on seedling emergence, then effects on weed seedling emergence will not be significantly related to a single factor unless the action of that factor dominates over the actions all the other factors or a set of dominant factors are collinear, and
c. If the soil physical (e.g., temperature, moisture, pH, and aeration) and chemical characteristics (e.g., concentrations of organic and inorganic compounds) in the presence and absence of cover crop tissues/residues on/in the field are not similar/

identical under laboratory conditions, then effects of tissues/residues on weed seedling emergence observed in the laboratory will not be representative of what happens in the field, but represent only potential effects that might occur under field conditions.

4.1.2.6 To Determine the Respective Importance of Shoot and Root Residues in Regulating Early Broadleaf Weed Seedling Emergence (Section 3.4.6)

a. If organic and inorganic compounds including phenolic acids are different in shoot and roots of glyphosate-desiccated cover crop residues, then their effects (e.g., inhibition/stimulation) on weed seedling emergence will also be different. Note: This difference is further enhanced by the difference in distribution of the shoots (concentrated at the soil surface) and roots (distributed through a soil volume), and
b. If leaching of glyphosate-desiccated shoot cover crop residues is required to stimulate/inhibit weed seedling emergence, then inadequate rainfall or rainfall at the wrong time will result in no observable effects on weed seedling emergence, no matter how stimulatory or inhibitory the shoot cover crop residues may be in laboratory studies. Note: In the event of inadequate or inappropriate rain fall events the effects of root residues will dominate as long as soil moisture is appropriate.

4.1.2.7 To Determine Under Controlled Conditions How Phenolic Acid Containing Plant Tissues/Residues Mixed into Soil Modify Phenolic Acid-Utilizing Bulk-Soil and Rhizosphere Microbial Populations (Section 3.4.7)

a. If (When) phenolic acid enrichment of seedling-microbe-soil systems simultaneously inhibits sensitive seedling emergence or growth and stimulates phenolic acid-utilizing microbes in the rhizosphere/rhizoplane of the inhibited seedlings, then phenolic acids at similar concentrations released from cover crop residues will bring about similar behavior for seedlings and rhizosphere/rhizoplane microbes. Note: It is possible that inhibition is due to other organic inhibitors in the residue instead of or including phenolic acids. We are also assuming that the ability of microbes to utilize phenolic acids cannot be selected/induced by other organic molecules. Finally, we are assuming that other organic and inorganic compounds released with the phenolic acids do not enhance, neutralize, or counter the effects of phenolic acids, and
b. If cover crop residues in field plots are associated with a stimulation of phenolic acid-utilizing microbes in the rhizosphere/rhizoplane, then phenolic acids from cover crop residues in the field plots have been transferred from the bulk soil to the rhizosphere/rhizoplane of the inhibited weed seedlings. Note: This is particularly true for aboveground shoot residues. However, I suppose one can argue that this could also happen if residues were located within the rhizosphere

(e.g., residues mixed into soil or root fragments located in the soil next to roots) in that case we can at least say that the phenolic acids entered the rhizosphere in sufficient quantities to have an effect on microbes.

4.2 Final Comments

To use circumstantial evidence derived from field and laboratory bioassays to make a case for allelopathy in any ecosystem requires both a system-oriented and a reductionist approach. The system-oriented approach describes the behavior of the system and suggests testable hypotheses for determining cause and effect. The reductionist approach is a way of testing the potential of the proposed (hypothesized) cause and effect relationships. In theory good predictive empirical field and laboratory data will be useful in providing circumstantial evidence for or against the presence of plant–plant allelopathic interactions. However, recall that predictions deduced for hypotheses of field and laboratory bioassays are either accepted or rejected. The acceptance or rejection of such predictions, unfortunately, does not prove or disprove the presence of allelopathic interactions. Thus, the role of plant–plant allelopathic interactions in wheat no-till crop systems or for that matter any other ecosystem (except in the most extreme cases), will always be disputable.

So what regulates dicotyledonous weed emergence in wheat no-till crop ecosystems? There is circumstantial evidence for all of the mechanisms proposed (e.g., physical barrier and shading associated with wheat residues, the modification of soil temperature and soil moisture by wheat residues, the reduction in soil disturbance and changes in compaction in no-till systems, the immobilization of nutrients, such as nitrogen, and the release of allelopathic compounds). Published results and results described in this book suggest that the answer to this question varies with wheat cultivar, wheat tissue, weed species, environment (e.g., timing and rates of rainfall events, amount of residue, and soil type), rates, types, amounts, and timing of organic and inorganic compounds released into the soil, management practice (e.g., time of desiccation of wheat), and time of year.

However, the broad question "So what regulates broadleaf weed emergence in wheat no-till crop ecosystems?" was not the central focus of the research presented in this book, nor was the question "Can phenolic acids inhibit broadleaf weed emergence, such as morningglory, prickly sida or pigweed?" They obviously can. The central questions actually were the following: "How likely are the necessary phenolic acid concentrations and environmental conditions present in wheat no-till crop systems for inhibition of broadleaf weed seedling emergence to occur?" and furthermore, "Do phenolic acids ever have a dominant role in regulating broadleaf weed emergence or are phenolic acids just one component of a larger promoter/modifier/inhibitor complex that regulates broadleaf weed seedling emergence in wheat no-till crop systems?"

In order to answer these questions we must, among other things, know what the dynamics of active phenolic acid concentrations are in soils, primarily in soil solutions surrounding broadleaf weed seeds during times of germination or surrounding seedling roots during their initial period of growth. We would also have to have similar information regarding active concentrations and functional behavior of promoters, modifiers (organic and inorganic), and/or other inhibitors present in soil, as well as the role of other physical and biotic modifiers (see Section 2.6.1). This turns out to be a very complex problem, since available/active concentrations of promoters, modifiers, and/or inhibitors in soils are dynamic, i.e., turnover rates (production and loss) can range from high to low depending on the amount, age, and distribution of cover crop residues, rainfall events, soil moisture, temperature, types of microbes present, density of roots, etc. Furthermore, the actions of individual promoters, modifiers, and/or inhibitors in soil may be independent, neutral, synergistic, additive, or antagonistic to the action of others present in the complex. In addition, the tools presently available to determine available/active phenolic acid concentrations and turnover rates of phenolic acids in the soil, for even a single phenolic acid, are inadequate (see Section 4.3). Note: In regards to organic modifiers, present evidence suggests that their action is indirect (e.g., shifting microbial utilization from active compounds to other carbon compounds so that concentrations of active compounds decline at a slower rate).

So what insight do the data collected from both the laboratory and field studies described in this book provide regarding our two central questions? It should be very evident at this point that clear definitive or absolute answers to the two central questions proposed are not possible. As pointed out previously, no matter how much research we do on plant–plant allelopathic interactions or for that matter any other plant–plant interactions, a certain amount of ambiguity will always be present. The best that can be done is to present a listing of the pros and cons so that readers can reach their own conclusions. Since Science progresses by attempting to falsify hypotheses and not by trying to prove them, the con arguments will be presented first and the counter pro arguments second.

4.2.1 How Likely Are the Necessary Phenolic Acid Concentrations and Environmental Conditions Present in Wheat No-Till Crop Systems for Inhibition of Broadleaf Weed Seedling Emergence to Occur?

4.2.1.1 Cons and Pros from Model System Studies (Chapter 2)

Con 1: Phenolic acids are rapidly taken up by roots and rapidly neutralized by sorption to cell walls, incorporated into lignin, glucosylated, metabolized, and/or sequestered into the vacuoles. Thus phenolic acids are primarily neutralized and retained by roots and only a small fraction ever reaches the leaves.

Pro 1: Phenolic acids do not need to be taken up by roots and translocated to the shoots (i.e., leaves) to be effective. All that is required is root membrane contact. The primary effect of phenolic acids on sensitive seedlings appears to be associated with root cell membrane perturbation. Membrane perturbation leads to a reduction in net water and net nutrient uptake and a host of secondary effects (e.g., reduction in transpiration, photosynthesis, allocation of carbon to roots, leaf expansion).

Con 2: Selection and induction of phenolic acids-utilizing microbes in soil occur readily when phenolic acids are present.

Pro 2: Selection and induction of phenolic acid-utilizing microbes are not evident unless high concentrations of phenolic acids are supplied to soil systems over time and environmental conditions are appropriate. However, what is really more important here is the activity of these microbes, i.e., utilization rates for phenolic acid (see #3).

Con 3: Microbes in the bulk soil, rhizosphere, and rhizoplane rapidly utilized phenolic acids as a carbon source. In most instances, the resulting breakdown products were short lived and much less inhibitory or neutral in their effects, thus soil solution concentrations are generally low. Accumulation of breakdown products requires an anaerobic environment.

Pro 3: Microbial activity in field soils is frequently limited by the soil environment. In fact, it has been suggested that in field soils large numbers of microorganisms spend most of their lives just surviving in quiescent or dormant stages. Lavelle and Spain (2001) suggest that bulk-soil bacteria, for example, may be active 3 years out of 30 years. Phenolic acid-utilizing microbes in the rhizosphere or on the rhizoplane, however, are very likely to be active as long as roots are functional. What we presently lack is information on distribution (i.e., frequency and density) of phenolic acid-utilizing microbes within the rhizosphere and on the rhizoplane. Not all microbial breakdown products are less inhibitory than the original phenolic acids, some are much more inhibitory (Liebl and Worsham 1983). There are also many anaerobic sites (e.g., soil and organic particles/aggregates) in soils where breakdown products can accumulate. In addition, the concentration of individual phenolic acids required for a given inhibition declined when the number of phenolic acids and/or other readily available organic compounds (e.g., glucose, phenylalanine) increased. Available organic compounds, including phenolic acids, are numerous in soils. Finally the rate of transport (mass flow) by "transpirational pull" of available phenolic acids to root surfaces can be faster than their utilization by rhizosphere/rhizoplane microbes.

Con 4: Phenolic acids were rapidly and irreversibly sorbed to soil particles and/or incorporated into the recalcitrant organic matter within the soil.

Pro: 4: Very rapid irreversible sorption occurs on initial addition of high concentrations of phenolic acids to soil; subsequently, however, sorption is very slow for cinnamic acid derivatives and almost non-existent for benzoic acid derivatives. For soil systems with continuous production of phenolic acids a semi-steady state will occur and irreversible sorption will thus be very small. That phenolic acids and soils can potentially reach a steady state has been demonstrated in the continuous-flow system.

Con 5: Roots and associated microbes rapidly modified nutrient/soil solution, and rhizosphere/rhizoplane pH. In case of Hoagland's solutions (contains only nitrate as N source) the pH shift was towards the neutral side. Effects of phenolic acids on seedling growth occurred best under acidic conditions and effects were reduced/lost under neutral or basic conditions.

Pro 5: True, effects of phenolic acids on seedlings occur most readily under acidic conditions. However, soils have substantial buffering capacity, and thus pH changes due to root and microbial activity are unlikely to occur at the bulk-soil levels over short time intervals. Changes in pH resulting from root and microbial activity do occur within the rhizosphere and on the rhizoplane (Heckman and Strick 1996; Rao et al. 2002; Ortas and Rowell 2004). Alkalization or acidification of the rhizosphere occurs for many species because of changing cation-anion uptake ratios, particularly uptake of nitrate or ammonium, respectively. In cowpea, for example, the rhizosphere is alkalinized in the dark and acidified with light exposure of the shoots even when supplied with nitrate (Marschner and Römheld 1983; Rao et al. 2002).

Con 6: High concentrations of individual phenolic acids were required for the inhibition of germination and seedling growth, much higher than is normally observed in field soils.

Pro 6: Soils contain complex mixtures of phenolic acids. Mixtures of phenolic acids can be inhibitory to seedling growth even when individual phenolic acids making up that mixture are well below their individual inhibitory concentrations. In fact to obtain and maintain a given percent inhibition of seedling growth concentrations of the individual phenolic acids making up the mixture of phenolic acids had to be reduced as the number of phenolic acids in the mixture was increased. Effects of individual phenolic acids in a mixture at low concentrations were additive and at higher concentrations partially antagonistic. Thus the effects of non-inhibitory, as well as, inhibitory concentrations of individual phenolic acids were accumulative.

Soils contain not only complex mixtures of phenolic acids, but also complex mixtures of other organic compounds. The concentration of phenolic acid required for a given percent inhibition of seedling growth declined when other more readily microbial-utilizable organic compounds were added, e.g., non-inhibitory concentrations of glucose acted as modifiers. On the other hand, other organic compounds at the right concentrations can also be inhibitory (e.g., methionine) and modify the overall inhibition in a synergistic, additive, or partial antagonistic manner. As with glucose, the presence of methionine decreased the concentration of a phenolic acid required for a given inhibition. Finally, the high concentrations required for inhibition in laboratory bioassays were a result of over-simplified and environment-rich (e.g., nutrients, moisture, pH, and active microbes) model systems that were (are) not representative of field systems (e.g., they have different physiochemical and biotic environments), and

Con 7: Effects of phenolic acids on seedling processes, such as nutrient and water uptake, transpiration, and leaf expansion were readily reversed once phenolic acid contact was lost/removed from root surfaces. Thus root contact with phenolic acids at inhibitory concentrations must be maintained for extended periods to generate extensive long term effects (e.g., smaller leaf areas, reductions in biomass). Furthermore, since effects of phenolic acids were local in nature and not systemic, a substantial portion of the root system has to be in contact with phenolic acids for an extended time period before long-term effects are observed. For example, inhibition of leaf expansion of cucumber seedlings was directly related to the portion of the root system in contact with phenolic acids.

Pro 7: Phenolic acid concentrations in no-till wheat soils were actually fairly stable over the growing seasons studied. Additional input of phenolic acids (and other organic and inorganic compounds) from surface desiccated cover crops would occur with each rainfall/irrigation event and would be highest shortly after desiccation, frequently a crucial time for weed seedling establishment in crop systems. Over longer time intervals the release of the more tightly bound phenolic acids, etc. would be associated with decomposition (i.e., fragmentation, faunal and microbial activity, and leaching). Accumulation of organic and inorganic compounds leached from shoot cover crop residues would accumulate primarily at or just below the soil surface. Young weed seedlings have comparatively small root systems that are located just below the surface of the soil, and thus when seedling roots do come in contact with phenolic acids the percent surface area of roots in contact with phenolic acids will be large. DeFrank and Putnam (1978) and Putnam et al. (1983) have demonstrated that placing residues on the soil surface can lead to a formation of an inhibition zone near the surface of the soil.

4.2.1.2 Cons and Pros from Field Studies (Chapter 3)

Con 1: Extraction of field soils yielded low concentrations (< 4 μg/g soil) of available ("free" and reversibly sorbed) individual phenolic acids. Phenolic acids were highest near the surface of the soil and declined rapidly with soil depth. Movement of phenolic acids in soil solutions to roots would occur primarily by capillary movement that tends to be slow. Movement of phenolic acids by gravitational water would be sporadic. The likelihood that a large proportion of a root system will be in contact with phenolic acids at inhibitory concentrations appears to be very small.

Pro 1: Concentrations observed in field soils are residual concentrations, i.e., what is left after soil fixation, microbial utilization, root uptake, and leaching. What is really needed are data on phenolic acid inputs and their subsequent distribution (output) to soil sinks (e.g., clays, organic matter, roots, and microbes). The importance and need for input and output data were demonstrated in the continuous-flow system when inhibition of seedlings occurred even when available phenolic acids could not be recovered by extractions from soils within the system (i.e., no evident residual available phenolic acid concentrations). Movement to roots can be dramatically expedited by mass flow regulated by "transpirational pull" (Blum 2006). Preferential flows of solutes in soils can also occur (Jardine et al. 1989, 1990).

Con 2: Phenolic acids were rapidly lost from surface wheat cover crop residues after desiccation and thus release of phenolic acids to soils was limited in time, roughly 3 or 4 weeks, for the cover crop residues tested. The rate of and time for depletion of phenolic acids from surface residues was determined largely by the extent and frequency of rainfall/irrigation events. For most of the growing season phenolic acids in soil extracts were, in fact, largely derived from older decomposed organic residues and/or a result of microbial activity on cover crop residues, which is limited by environmental constraints.

Pro 2: Inhibition of weed seedling emergence, when it occurred, was consistently observed at the beginning of the growing season, after which weed seedling emergence recovered to various extents over time. This is in step with the declining phenolic acid concentrations in wheat cover crop residues. Similar observations have been made for hydroxamic acids in rye (Yenish et al. 1995). Inhibition early in the growing season can be highly beneficial to crop production since such delays in emergence or growth of weeds provides an opportunity for the crops to dominate the canopy and further suppress the weeds.

Con 3: No significant relationships between weed seedling emergence and soil phenolic acid content were observed. Concentrations of phenolic acids in no-till wheat soils were slightly higher than in no cover crop no-till soils, but these differences were comparatively small. Concentrations of phenolic acids in clover cover crop residues and their associated soils were higher than the small grain cover crop residues and their associated soils, yet clover cover crop residues/soils frequently stimulated weed seedling emergence.

Pro 3: Phenolic acids in bulk soil do not necessarily represent what actually contacts root surfaces since they provides no insight into the proximity of roots to phenolic acids and/or the movement/mass flow of phenolic acids in soils; thus significant relationships are unlikely to be observed from bulk-soil data. Besides, with the large number of phenolic acids in soil the concentration of any one individual can be very small, but the total of all the phenolic acids may still be inhibitory. The situation becomes even more complex when we add the role of other organic and inorganic compounds in the soil. See Section 4.3 for additional details. In regard to the clover phenolic acid data, total phenolic acid content of small grain and clover residues should not be compared directly since the Folin & Ciocalteu's phenol reagent is also reduced by a number of other organic and inorganic compounds, including some amino acids that are much higher in clover residues. As to the clover crop residues, the phenolic acid content in the clover residues declined much more rapidly over the first month than the small grain residues. Inhibitory effects on weed seedling emergence by clover residues occurred within the first 2–3 weeks after desiccation. The inhibitory effects of wheat residues for example, depending on species, lasted much longer. An additional aspect is the potentially higher microbial utilization rates of organic compounds in soils associated with clover residues that have much higher nitrogen levels.

Con 4: Total phenolic acid content in soils was overestimated by the Folin & Ciocalteu's phenol reagent because a number of other organic and inorganic substances also reduce the phenol reagent. In fact, some of these other organic compounds may have been important, if not more important, in the inhibition observed for soil extracts.

Pro 4: Total phenolic acid concentrations estimated by water-autoclave extraction and Folin & Ciocalteu's phenol reagent are not absolute values and thus should not be viewed as such, they are relative values. The following suggest that these relative values for total phenolic acid content can actually be useful and meaningful:

a. total phenolic acid concentrations in the soil extracts were correlated with concentrations of individual phenolic acids and the sum of 7 common simple phenolic acids in soil extracts,
b. soil extracts from wheat stubble tilled under/soybean and fallow/soybean systems (conventional tillage) were not inhibitory while soil extracts from wheat stubble/soybean systems (no-till) were inhibitory to morningglory and crimson clover, and
c. total phenolic acid content of soil extracts from wheat stubble/soybean systems was inversely related to the inhibition slopes (biological activity of extracts) of radicle/hypocotyl lengths of crimson clover.

Con 5: Inhibition of weed seedlings is also determined by a variety of other factors (e.g., pH, soil moisture, soil temperature, and solute potential) besides phenolic acids. It is not entirely clear as to which factor or factors dominate in field soils.

Pro 5: The following observations are all suggestive and provide hope that the potential roles of environmental factors and the effects of phenolic acids can eventually be differentiated for no-till wheat/crop systems:

a. effects of phenolic acid mixtures, pH, and solute potentials of simulated (model) wheat stubble/soybean (no till) soil extracts were independent from each other,
b. total phenolic acid content of soil extracts from wheat stubble/soybean (no-till) systems was inversely related to the inhibition slopes (biological activity of extracts) of radicle/hypocotyl lengths of crimson clover, and
c. the relationship observed in "b." was not observed for wheat stubble tilled under/soybean and fallow/soybean (conventional-till) soil extracts, and

Con 6: Very inconsistent effects ranging from inhibition to stimulation were observed even when phenolic acid content of soils was similar within and among the various growing seasons. For example, during the 1996 and 1997 seasons, wheat root residues stimulated morningglory and prickly sida seedling emergence, while shoot (surface) residues had no effect on seedling emergence. When both wheat shoot and root residues were present seedling emergence for these two weed species ranged from no effect to stimulation for the 1996 and 1997 seasons, but inhibition for the 1992 and 1993 seasons.

Pro 6: This is not entirely surprising since rainfall/irrigation and soil moisture varied considerably from growing season to growing season and adequate rainfall/irrigation and soil moisture are required for the release and availability of water soluble substances, such as phenolic acids. There are also additional factors, e.g., nitrogen, pH, residue chemistry, residue age, and microbial populations, which enhance or eliminate the effects of desiccated cover crop residues on weed seedling emergence. Very sensitive species, such as pigweed, however, appear to be much more consistently suppressed.

4.2.2 Do Phenolic Acids Have a Dominant Role in Regulating Broadleaf Weed Seedling Emergence or Are Phenolic Acids Just One Component of a Larger Promoter/Modifier/Inhibitor Complex that Regulates Broadleaf Weed Seedling Emergence in Wheat No-Till Crop Systems?

If you have concluded at this point that phenolic acids are unlikely (i.e., accepted the con arguments) to function as inhibitors of broadleaf weed seedling emergence in wheat no-till crop systems, then this question is immaterial. If you have concluded that phenolic acids are likely to, may have, or do have a role in inhibiting broadleaf weed seedling emergence (i.e., accepted the pro arguments), then the following comments are pertinent.

Given the complex nature of soil processes (e.g., sorption, microbial utilization and synthesis, root uptake, and leaching) and of organic and inorganic compounds in soil, given that any of these compounds may be stimulatory, neutral or inhibitory depending on their concentrations and physicochemical states, given that the actions of individual compounds in mixtures of compounds can be synergistic, additive, or antagonistic, given that changes in the physicochemical environment (e.g., pH, nutrient status, moisture, and temperature) can modify the behavior of soil processes and actions and interactions of organic compounds, and given that the sensitivity of weed species to organic compounds varies with taxon, age of seedling, and state of seedling acclimation and "health" no all-encompassing or definitive statement can be made as to whether or when phenolic acids are dominant inhibitors or whether they are simply components in promoter/modifier/inhibitor complexes in wheat no-till crop systems. It really depends on the circumstances prevalent at the time. If pressed for an answer I would hypothesize that in most instances phenolic acids are unlikely to have a dominant role in regulating broadleaf weed seedling emergence in wheat no-till crop systems. Given the present data on phenolic acids in cover crop residues, their release rates, and available concentrations in soils I can come up with a number of cons for a dominant role, but not really any convincing pros. However, I suspect this will not be the final word on this subject. Below the reader will find a listing of the cons for a dominant role and by default pros for phenolic acids being simply one component of a promoter/modifier/inhibitor complex in wheat no-till crop systems.

Con 1: A number of allelopathic agents besides phenolic acids have been identified in wheat tissue/residues, for example, organic acids, hydroxamic acids, and volatile compounds (Patrick 1971; Lynch 1977; Willard and Penner 1976; Tang and Waiss 1978; Buttery et al. 1985; Shilling et al. 1985; Niemeyer et al. 1989; Blum et al. 1991; Wu et al. 2001; Macías et al. 2005; Mathiassen et al. 2006). A number of these are just as or more phytotoxic than simple phenolic acids.

Con 2: Usually within hours to days after glyphosate desiccation cell membranes are disrupted and all the soluble and readily available organic and inorganic compounds (including phenolic acids) in the dead cells are released into the soil by leaching during the first adequate rainfall event. The most dramatic effects on weed seedling emergence will take place at that point in time assuming seeds are germinating or seedlings are present. Subsequently the release of bound organic and inorganic compounds from the remaining wheat residues will be at much slower rates and lower concentrations (i.e., chronic releases) by leaching of faunal and microbial breakdown products (i.e., decomposition) as long as environmental conditions are appropriate. Effects of phenolic acids in soils are not independent of other organic and inorganic compounds present.

Con 3: Organic and inorganic compounds (including phenolic acids) leached from wheat residues ("new") are added to organic and inorganic compounds (including phenolic acids) already in the soil and soil solution ("old"; Flaig 1971; Paul and Clark 1989; Lavelle and Spain 2001; Strobel 2001). The action and interactions of the resulting combinations and concentrations of "old" and "new" organic and inorganic compounds (including "new" and "old" phenolic acids) on broadleaf weed seedling processes will vary depending on how the residue leachates have modified the concentrations of all of the active ("old" and "new") compounds present in the soil and the physicochemical environment of the soil (Rasmussen and Einhellig 1977; Einhellig et al. 1982; Blum et al. 1985a, 1989, 1993; Gerig and Blum 1991, 1993; Pue et al. 1995; Blum 1996).

Con 4: Once organic compounds enter the soil solution soil processes reduce the available/active fractions of the organic compounds, including phenolic acids, by reversible and irreversible sorption, polymerization, oxidation-reduction, leaching, and microbial utilization (Huang et al. 1977, 1999; Dao 1987; Dalton 1999; Schmidt and Ley 1999; Ohno 2001; Blum 2004, 2006). Active concentrations of organic compounds in soil solutions, therefore, tend to be relatively low. Note: There is, however, some uncertainty about the actual functional concentrations of organic compounds (including phenolic acids) in soil solutions because of the lack of reliable data on the actual flows of compounds into and out of soil solutions. For example, there is considerable evidence that as organic compounds are lost from the soil solution they are replaced by organic compounds reversibly sorbed to soil particles, root exudates and secretions, lyses of root cells, microbial synthesis, and decomposition and leaching of organic residues, etc. Finally, presence of low non-inhibitory individual concentrations can still result in inhibition if the action of the sum of all the low non-inhibitory concentrations is inhibitory.

Con 5: The primary effects of phenolic acids are localized to the initial site of root contact since seedling processes were directly related to the percent of their root systems in contact with phenolic acids (Klein and Blum 1990; Lyu and Blum 1990; Lehman et al. 1994; Lehman and Blum 1999) and were poorly related to phenolic acids taken up by roots (Lehman and Blum 1999). There is, therefore, little data to suggest that the effects of phenolic acids are systemic in nature. Furthermore, once phenolic acid contact is lost seedling processes recover (Blum and Dalton 1985; Blum and Rebbeck 1989), and

Con 6: The primary effect of phenolic acids appears to be membrane perturbation (Glass 1973, 1974, 1975; Glass and Dunlop 1974). Thus phenolic acids must reach cell membranes to have an effect. The phenolic acids reaching cell membranes from soil solutions surrounding roots are reduced/moderated/regulated by the bulk-soil, rhizosphere, and rhizoplane microbes, sorption to primary and secondary cell walls, and feedback regulation by seedlings (Harborne 1982; Shann and Blum 1987; Blum and Schafer 1988; Fry 1988; Shafer and Blum 1991; Blum 1998; Blum et al. 2000; Blum and Gerig 2005).

It appears to me that given the complexity of soil systems and seedling responses to the considerable number of organic and inorganic compounds already in the soil and the soil solution that the addition of phenolic acids (including a range of other potential identified phytotoxins and modifiers) released from wheat residues is unlikely to lead to a dominant role of phenolic acids in regulating weed seedling emergence in wheat no-till crop systems. A role in the action of a promoter/modifier/inhibitor complex within the soil solution is much more likely.

4.3 The Present Paradigm

Among the criteria listed in Chapter 1 to establish plant–plant allelopathic interactions were the following:

a. The affected plants must come in contact with and interact with the organic promoters or inhibitors produced directly or indirectly (e.g., modified by microorganisms) by an allelopathic plant, and
b. These organic promoters and inhibitors must be at sufficient concentrations and be present for a sufficient length of time to modify plant function and growth of the receiving plant.

It should be clearly evident at this point that demonstrating these two criteria in nature is quite a challenge, a challenge that has not been fully met at this writing. Our inability to clearly demonstrate these criteria for plant–plant allelopathic interactions in nature is the product of an inappropriate paradigm and the lack of adequate tools. For the sake of clarity and brevity I will limit my comments below to inhibitory plant–plant allelopathic interactions.

Throughout the allelopathic literature there is this general impression that a few simple organic inhibitors can be identified as the causative agents for plant–plant allelopathic interactions. No one says that directly, it is just implied by the focus of the research or by the way the research is presented. It appears thus to be an informal mind set, or if you prefer an informal paradigm, a "herbicide paradigm". I suspect in the majority of cases this paradigm is not true, it clearly appears not to be true for phenolic acids. In fact, inhibition is a result of a complex set of interactions between the types of plants involved, the ever changing ratios of organic and inorganic promoters, modifiers, and inhibitors in the soil and soil solution and more importantly in the rhizosphere/rhizoplane, the physicochemical environment, and the types and activities of micro, meta, and macro organisms present. Note: Inhibitors in the present case must dominate over promoters. Even if we ignore everything but available/active organic compounds in soil solutions, the complexity is awe-inspiring. After all, organic compounds in soil solutions are very complex, being composed of not only phenolic acids, but also amino acids, various organic acids, carbohydrates, nucleotides, and vitamins, just to name a few (Flaig 1971; Paul and Clark 1989; Lavelle and Spain 2001; Strobel 2001; Blum 2006). Any one of these organic compounds, at the right concentration and physicochemical state (e.g., protonated or ionized), could be inhibitory, while at another concentration and physicochemical state could be neutral or stimulatory to the plant process being measured (Duke et al. 2006; Belz et al. 2007; Belz 2008). Furthermore, as organic compounds are removed from the soil solution by root uptake, microbial utilization, oxidation-reduction, sorption, etc. they can be replaced by reversibly sorbed compounds on soil particles, root exudates and secretions, lyses of root cells, microbial synthesis, and decomposition and leaching of organic residues, etc. Clearly, all these soil processes and their interactions are much too complex to limit our focus on just a few simple inhibitors (i.e., silver bullets) as causative agents, unless they are extremely toxic. A number of the natural products that have been identified as allelopathic agents, including phenolic acids, tend not to be extremely toxic or even highly toxic.

4.3.1 Phenolic Acids in Soils: Soil Extractions and Dose Response

The two methods most frequently used to establish that inhibited plants are in contact with and interact with organic inhibitors produced by putative allelopathic plants or plant residues and to establish that these inhibitors are present at sufficient concentrations and for sufficient lengths of time are soil extractions and dose response studies. You may have noticed that this was the approach taken for the research presented in this book. There are, however, now several aspects of this approach that are troubling.

For simplicity let us continue to use soil phenolic acids as our model inhibitors, recognizing that similar statements can probably be made for all promoters, organic and inorganic modifiers, and inhibitors in the soil. The following critiques regarding the usefulness of soil extractions and dose response studies are specifically limited to answering the following question: "Can we demonstrate by these procedures that phenolic acids, or for that matter other allelochemicals, actually inhibit plant growth and development under field conditions?"

a. Soil Extractions: Unfortunately, there are presently no combinations of extraction and quantification procedures that can provide absolute values for all the available or more importantly active phenolic acids in the bulk soil or rhizosphere/rhizoplane. Why not?

    1. Phenolic acids in soil solutions, reversibly sorbed on soil particles, and on and in plant residues range from simple phenolic acids (e.g., ferulic acid) to complex polymers, such as tannins. Soils and residues also contain a variety of polymers that contain phenolic acid moieties, e.g., humic acid, fulvic

acid, lignin, and melanins. Each of these phenolic acids or phenolic acid moieties are held at different strengths within soil (on clays or on and in organic matter) and plant residues (on or in) and are released at different times and rates to soil solutions. Different types of extraction procedures would have to be used at the appropriate times to determine the potential available/active fractions being released into the soil solution. This assumes we knew what fractions should be extracted, how they should be extracted, and when they should be extracted.

2. Even if we could extract all of the available/active phenolic acids with a single extraction procedure their different physical and chemical properties and their large number would make it essentially impossible to identify and quantify them all.
3. Soil extractions cannot differentiate between sources of phenolic acids, e.g., are they from shoots or roots of the putative allelopathic plant or the receiving plant, are they from plant residues on or in the soil, are they from faunal and microbial activity on new or old organic residues in the soil, etc.
4. Extractions of bulk soil are very likely to be of little value in determining inhibition of phenolic acids unless the soil's available phenolic acid content is closely related to that in the rhizosphere/rhizoplane or the soil solution surrounding or in proximity to roots. Phenolic acid inhibition occurs at the rhizosphere/rhizoplane level not at the bulk-soil level.
5. Soil extractions under the best of circumstances provide information on net or residual available phenolic acids but there is no reliable way, at present, to determine when or where these compounds function as active inhibitors, and
6. Many inhibitory effects of phenolic acids are rapidly reversed once phenolic acids are eliminated (e.g., leaching, microbial activity, root uptake) from the root environment. Sustained inhibition requires that phenolic acids at the right concentration and physicochemical state, and under the right environmental conditions be continuously present and in contact with a large fraction of the root system. Unfortunately available/active phenolic acids in soils are dynamic, ever changing. Realistic data on flows into and out of various phenolic acid pools in the soil are notoriously difficult to get by soil extractions, since each extraction represents a point in time and, if for example, input equals output, then net flow between two extraction times will appear to be zero. In fact, considerable flows may be occurring between sources and sinks within the soil that are just not detectable by way of soil extractions.

Ultimately, for soil extracts to be biologically meaningful requires that soil extracts be representative of soil solutions not just at one point in time but over time. How to achieve that is a major conundrum/mystery for researchers studying plant–plant allelopathic interactions!

b. Dose Response: Dose response studies are only useful in identifying phenolic acid inhibitors in field soil if the right combination of phenolic acids and the right range of concentrations are used and the dose response studies are carried out under the right environmental conditions (e.g., appropriate concentrations, levels, or populations of other organic and inorganic compounds, pH, water potential, temperature, aeration, and microbial populations). Why?

1. Theoretically all phenolic acids are stimulatory, neutral, or inhibitory depending on their concentration and their physicochemical state. This is true for other soluble organic and inorganic compounds as well, and
2. It has been determined that physicochemical state (e.g., protonated or ionized) and concentration are important indicators of phenolic acid inhibition. We also now know that the presence of other organic and inorganic compounds, the number and type of phenolic acids present, changing phenolic acid concentrations, soil type, microbial activity, state of acclimation and "health" of receiving plant, proportion of the root system in contact with phenolic acids, presence of other stressors, aeration, soil moisture and water potential, rhizosphere/rhizoplane pH, etc. all can substantially modify the inhibitory activity of phenolic acids. The modifications can range from enhancing to eliminating the inhibitory effects of phenolic acids. Thus inhibition based simply on concentration and physicochemical state of phenolic acids turn out to be poor predictors of actual role of phenolic acids under field conditions. Unfortunately most dose response studies in the literature turn out to be just that.

How to design biologically meaningful dose response studies that are representative of field conditions is thus another major conundrum/mystery for researchers studying plant–plant allelopathic interactions! This is not to say that extractions and dose response studies are not useful tools, they are, but to say that they have their limits when it comes to providing realistic and acceptable evidence that plant–plant allelopathic interactions are occurring in the field.

4.4 A Modified Paradigm

What is required is a more "holistic" approach to the study of plant–plant allelopathic interactions. What we need is a modified paradigm. The "herbicide paradigm" is too limiting. We need a paradigm that emphasizes promoter/modifiers/inhibitor complexes of soil solutions, the interactions of available organic (including phenolic acids) and inorganic compounds in the soil solutions with each other and with their immediate environment, and the role of the rhizosphere/rhizoplane. We need a paradigm that places less emphasis on potential individual inhibitors, i.e., "silver bullets" and bulk-soil characteristics.

Why is this modified paradigm necessary?

a. Observed inhibition of sensitive seed or seedling is a result of a complex set of interactions between promoters, modifiers, inhibitors, stressors, and environmental factors.

b. Complexes are very dynamic in behavior since any compound (organic or inorganic) in the complex (mixture) may be inhibitory, neutral, or stimulatory depending on its concentration, environment, sensitivity of the species, etc. and
c. Effects of individual components within a complex and the immediate environment may be independent, additive, synergistic, or antagonistic to each other.

4.4.1 Criteria for Plant–Plant Allelopathic Interactions: An Update

From a theoretical standpoint the criteria provided in Chapter 1 of this book have not changed; what has changed is the setting or landscape of these criteria. Associated with each criterion the reader will find a note that will either propose no change, suggest a clarification, or updates the criterion to be consistent with a "holistic paradigm".

a. Patterns of stimulation or inhibition of plants on other plants must be shown. Note: No change proposed for this criterion.
b. The observed patterns cannot be solely explained by physicochemical modifications of the environment (other than promoters and/or inhibitors), utilization of substances as a source of nutrients, carbon and/or energy, transfer through mycorrhizae and/or root grafts (please note ambiguity in Chapter 1), and/or biotic factors such as resource competition, herbivory, or disease. Note: The focus of this criterion will still be on "cannot solely be explained by" and thus that aspect of this criterion does not change. However, the roles of the physicochemical and biotic environments must not be viewed simply as causative agents but must also be viewed as modifiers of causative agents since for example pH, soil moisture, organic and inorganic compounds, aeration, microbial populations, etc. will all influence the final outcome by changing the active concentrations of promoters, modifiers, and/or inhibitors.
c. The putative allelopathic plants or their residues must produce/contain and release organic substances into the environment that will ultimately be capable of stimulating or inhibiting the function or growth of associated plants. Note: More emphasis must be given to what is released from plant tissues/residues to the soil and their subsequent degradation products in the soil and not what is contained in plant tissues/residues. Concentrations of promoters, modifiers, and inhibitors within plant tissues/residues are not representative of what is actually released from tissues/residues over time to the soil or their subsequent degradation products in the soil. Thus plant tissue/residue or their extracts are unlikely to be representative of promoter/modifier/inhibitor complexes in soils acting on seeds or roots.
d. The affected plants must come in contact with and interact with the organic promoters or inhibitors produced directly or indirectly (e.g., modified by

microorganisms) by an allelopathic plant. Note: Updated version – The affected plants must come in contact with and interact with a complex of organic compounds produced directly or indirectly (e.g., modified by microorganisms) by putative allelopathic plants. A preponderance of these organic compounds must function as promoters, and/or inhibitors alongside all the other organic and inorganic compounds (modifiers) in the soil in such a way that the sum of their actions will be either stimulatory or inhibitory to receiving plants, and

e. These organic promoters and inhibitors must be at sufficient concentrations and be present for sufficient length of time to modify plant function and growth of receiving plants either directly (e.g., impact on root membranes and/or cell process) or indirectly (e.g., impact on nodule or mycorrhizae formation, development, and/or function). Note: Updated version – These promoter/modifier/inhibitor complexes dominated by either organic promoters or inhibitors must be present at sufficient level of activity and time to modify plant function and growth of receiving plants either directly (e.g., impact on root membranes and/or cell process) or indirectly (e.g., impact on nodule or mycorrhizae formation, development, and/or function).

4.4.2 Potential Tools

You say, easier said than done. I would agree. Clearly new tools/techniques need to be found and developed consistent with a "holistic" approach. However, there are some potential tools/techniques that are already available and may prove to be useful.

a. Blum et al. (2000) found that addition of equal-molar mixtures of phenolic acids to soil-microbe-seedling systems resulted in the simultaneous inhibition of seedling growth and a stimulation of bacteria in the rhizosphere that could utilize phenolic acids as a sole carbon source. Staman et al. (2001) using the same soil-microbe-seedling system also found this inverse relationship when wheat or sunflower plant tissues were added instead of phenolic acids. However, in the latter case this inverse relationship was inconsistently expressed between experiments and over time. Based on these observations Staman et al. proposed the following: If under field conditions plant residues containing phenolic acids inhibit sensitive seedlings and stimulate phenolic acid utilizing microbes in the rhizosphere of those inhibited seedlings, then phenolic acids released from these plant residues have reached the rhizosphere/rhizoplane of the inhibited seedlings and are wholly or partly responsible for the stimulation/inhibition observed. For additional scenarios see Section 3.5.5.2.

   I would suggest that given the right conditions and bioassay species this approach may also turn out to be useful not only for phenolic acids but also for other suspected allelopathic agents. The benefit of this approach is that the relationships between seedling inhibition and stimulation of phenolic acid-utilizing rhizosphere bacteria are determined in the presence of a complex of organic and inorganic compounds, i.e., plant residues, representing a promoter/modifier/inhibitor complex. On the other hand one might conclude that this approach is not far removed from the "herbicide paradigm". I would agree; however it is a step in the right direction.

b. Weidenhamer and collaborators used polydimethylsiloxane (PDMS) coated stir bars and optical fibers and PDMS tubing (i.e., probes) located within the root zone/rhizosphere of plants grown in the greenhouse and/or in the field to capture and quantify allelochemicals released by roots. PDMS is a nonpolar sorptive material in which analytes dissolve and from which analytes can readily be desorbed for analysis. Since samples can be taken over time relative estimates of fluxes of compounds within the rhizosphere (e.g., sorgoleone for sorgum roots, thiophenes from marigold roots) can be determined (Weidenhamer 2005, 2007; Mohney et al. 2009). Weidenhamer (2005, 2007) and Mohney et al. (2009) suggest that PDMS probes can be used to determine nonpolar, but also possibly some polar, allelochemics in the rhizosphere. PDMS fibers have also been inserted into the stems of tomato, common ragweed and purslane to detect their uptake of allelopathic agents such as cineole, camphor, menthol, and/or coumarin from soil (Loi et al. 2008). Thus PDMS probes appear to be extremely useful in monitoring allelopathic agents exudated, secreted, and/or released by cell lyses into the rhizosphere and taken up by roots from the soil. In addition, PDMS probes are sensitive to low concentrations, are easily installed, are inexpensive, and can be reutilized. PDMS bars and optical fibers would have to be inserted and removed from the root zone/rhizosphere for time studies and thus would be disruptive to the root system under study. PDMS tubes, however, could be placed into the soil at planting and could be sampled nondestructively on a daily basis. The utilization of PDMS tubes/fibers is consistent with the modified paradigm in that they provide a tool by which we can obtain data on the transfer of compounds from roots to the rhizosphere/bulk soil or from the bulk soil/rhizosphere into the plant roots in the presence of other organic and inorganic compounds.

c. Blum et al. (1992) used serial dilutions of water-autoclave soil extracts from wheat stubble/soybean (no-till) soils in petri dish germination bioassays to determine biological activity of soil extracts. Solute potential, pH, total phenolic acid, and ion content were determined for each undiluted extract. Resulting slopes (biological activity) of wheat stubble/soybean soil extracts were then regressed against undiluted soil extract characteristics to determine the potential role of each characteristic, i.e., solute potential, pH, total phenolic acid, and ion content. This approach has also been found useful for simulated (model) soil extracts (Blum et al. 1992). Yes, soil extracts have their limitations (e.g., there is no guarantee that soil extracts are representative of ambient soil solutions; see Section 4.3), but even with their limitations bioassays based on soil extracts (promoter/modifier/inhibitor complexes) will/should be more meaningful than bioassays on individual putative phytotoxic compounds.

d. Dalton (1993, 1999) suggested that particularly in wet mesic soils lysimeters (porous ceramic cup vacuum samplers) may be used to collect soil solutions.

Upon testing the feasibility of using lysimeters in the field and in the laboratory he found that when lysimeters were properly preconditioned that phenolic acids captured by porous clay cups were representative of the phenolic acids in soil solutions surrounding these cups. Others have made similar observations for nutrient ions (Grover and Lamborn 1970; Neary and Tomassini 1985; Debyle et al. 1988; Hughes and Reynolds 1988). I would hypothesize that this will also be true for other organic compounds and thus might be a means of collecting the whole or at least representative soil solution samples of promoter/modifier/inhibitor complexes (organic and inorganic). Once collected, soil solutions could be rapidly filter sterilized to retain their chemical properties for subsequent bioassays to determine their biological activity (inhibition slopes). Resulting biological activity regressed against soil solution characteristics collected might be a step in providing more "realistic" evidence for allelopathic plant–plant interactions. This of course assumes that the necessary controls are included (e.g., pH, water potential, inorganic ions) in the bioassays. For a potential use of lysimeters in less mesic soils see Blum (2004).

Soil solutions may also be collected by centrifugation from moist soil samples (Zabowski 1989; Zabowski and Ugolini 1990; Giesler et al. 1996; Strobel 2001). However, the amount obtained by centrifugation may only be sufficient for chemical analyses, i.e., too small for bioassays. Testing soil solutions, however, is only the first step. We also need quantitative assessments of how individual organic and inorganic compounds and physicochemical factors (e.g., pH, temperature, concentrations) of soil solutions interact that bring about inhibitory effects. This might be accomplished with simulated (model) soil solutions based on the chemical analyses of solutions collected by centrifugation.

e. Bioassays with or without activated carbon to bind allelopathic agents in soil can be useful in identifying the presence of potential allelopathic interactions as well as separating inhibitory plant–plant allelopathic interactions from resource competition (Mahall and Callaway 1992; Nilsson 1994; Wu et al. 2000; Ridenour and Callaway 2001; Bertin et al. 2009). Since activated carbon, depending on its source and environment, will adsorb a range of both organic and inorganic compounds (e.g., promoters, modifiers, and inhibitors) from soil solutions (Mattson and Mark 1971; Cheremisinoff and Ellerbusch 1978) utilizing activated carbon is consistent with the "holistic" approach. However, care must be taken in reaching conclusions in regard to actual cause and effect when activated carbon is utilized since the addition of activated carbon to soil may also modify soil characteristics such as moisture retention, aeration, and nutrition, and
f. The inclusion of a range of densities in bioassays can also be very useful. Under the right experimental conditions density-dependent effects may be used to identify potential inhibitory allelopathic interactions and distinguish inhibitory plant–plant allelopathic interactions from resource competition (Weidenhamer et al. 1989; Thijs et al. 1994; Weidenhamer 1996) as long as autotoxicity and/or hormesis are absent (An et al. 1993; Belz et al. 2005; Sinkkonen 2001, 2003, 2007). Density bioassays are clearly consistent with a "holistic" approach.

4.5 Concluding Remarks

A central tenet of plant–plant allelopathic interactions studies has been and continues to be that a single inhibitor or a group of similar inhibitors can be isolated, identified, and quantified in plant tissues/residues and field soils and that based on subsequent dose response studies of the inhibitors identified conclusions can be drawn regarding the presence or absence of plant–plant allelopathic interactions (i.e., the "herbicide paradigm") in the field. The benefit of this approach is that there is a clear direction to the research, e.g., defining the roles and activities of the target compound or similar compounds. It is also consistent with the heuristic guideline that states that the simplest explanation or strategy tends to be the best ones (Occam's razor). Theoretically this appears straight forward, but in practice providing acceptable evidence for plant–plant allelopathic interactions even for a single compound turns out to be very difficult, unless the compound is extremely toxic.

The research presented in this book, clearly demonstrates that things are far more subtle and complex than the "herbicide paradigm" would imply, and that the focus on a single compound or a group of similar compounds is somewhat misdirected, particularly if the compound or compounds in question have a low phytotoxicity. The proposed "holistic paradigm" shifts the focus away from a target compound or compounds to the soil and soil solution surrounding roots. Soils and soil solutions unfortunately are dynamic and composed of numerous organic and inorganic compounds each with different solubilities, mobilities, concentrations, physicochemical states, activities, and/or actions (neutral, or act as promoters, modifiers, or inhibitors) on plant roots. From a research stand point the focus on soil solutions is not very appealing. What is, therefore, needed is research that will determine the fewest elements of soils and soil solutions required to identify and characterize plant–plant allelopathic interactions under field conditions.

I do not want to leave the reader with the perception that the two paradigms are incompatible, they are not since the "herbicide paradigm" is a subunit of the "holistic paradigm". In fact the theoretical seedling-microbe-soil model system presented at the beginning of this book (Fig. 2.1) does not need to be changed to accommodate both paradigms. Furthermore, we have made considerable progress in understanding what elements must be accounted for in soil and soil solution bioassays, these include solute potential, pH, aeration, solar radiation, temperature, stressors, organic and inorganic compounds, microbial populations in the bulk soil and rhizosphere/rhizoplane, and seedling acclimation, "health", growth rates, and root distribution. Researchers can account for many of these elements by having the necessary controls and treatments included in their bioassays. You say that is a tall order, I would agree. Future research on the relationships and importance of these elements could help to reduce the number of elements that must be included in bioassays consistent with Occam's guidelines. For example, co-linear elements could be identified by using various statistical tools and/or principal components analysis (Blum et al. 1991; SAS Institute Inc. 1999) and once identified a proxy element could be identified and substituted for these co-linear elements. Total phenolic acid based on the Folin & Ciocalteu's phenol reagent may be such a proxy since the phenol reagent is not only reduced by monohydric and polyhydric phenolic substances but also by other organic and inorganic compounds and reducing agents (McAllister 1969; Box 1983).

I hope that it is now clearly evident that the use of laboratory and field bioassays of residues, extracts, or individual compounds without the appropriate range of physicochemical and biotic elements (e.g., concentrations, solute potentials, pH values, nutrition, temperature, etc.) and adequate/realistic controls are no longer acceptable when attempting to identify/characterize the existence of plant–plant allelopathic interactions in nature. In spite of the difficulties a much more "holistic" approach is now required both in the laboratory and in the field. We need a renewed effort to develop tools and procedures that isolate and then reintegrate the actions of physicochemical and biotic elements of soil solutions, rhizospheres, and rhizoplanes that regulate/modulate allelopathic plant–plant interactions.

Finally, I would be remiss if I did not point out that the emphasis in this book has been on phenolic acids, their interactions with other physical, chemical, and biotic components within the soil, and their potential role in plant–plant allelopathic interactions for no-till cover crop systems, primarily wheat cover crops. There are, however, a number of additional aspects regarding allelopathic weed emergence/control by cover crop residues that have only been mentioned indirectly or in passing. For example, the role of other potential phytotoxins such as hydroxamic acids (Willard and Penner 1976; Niemeyer et al. 1989; Wu et al. 2001; Macías et al. 2005; Mathiassen et al. 2006) and volatile compounds (Buttery et al. 1985; Bradow and Connick 1990; Bradow 1991), or the production of phytotoxins by microbes that flourish on specific types of cover crop residues or in general within the soil (DeFrank and Putnam 1985; Heisey et al. 1985, 1988). All of these add to the urgency to shift our research focus from a "herbicide paradigm" to a "holistic paradigm".

References

An M, Johnson IR, Lovett J (1993) Mathematical modeling of allelopathy: biological response to allelochemicals and its interpretation. J Chem Ecol 19:2379–2388

Belz RG (2008) Stimulation versus inhibition – bioactivity of parthenin, a phytochemical from *Parthenium hysterophorus* L. Int Dose-Response Soc 6:80–96

Belz RG, Hurle K, Duke SO (2005) Dose response – a challenge for allelopathy? Nonlin Biol Toxicol Med 3:173–211

Belz RG, Velini ED, Duke SO (2007) Dose/response relationships in allelopathy research. In: Fujii Y, Hiradate S (eds) Allelopathy: new concepts and methodologies. Science Publishers, Enfield, NY, pp 3–29

Bertin C, Harmon R, Akaogi M, Weidenhamer JD, Weston LA (2009) Assessment of the phytotoxic potential of *m*-tyrosine in laboratory soil bioassays. J Chem Ecol 35:1288–1294

Blum U (1996) Allelopathic interactions involving phenolic acids. J Nematol 28:259–267

Blum U (1998) Effects of microbial utilization of phenolic acids and their phenolic acid breakdown products on allelopathic interactions. J Chem Ecol 24:685–708

Blum U (2004) Fate of phenolic allelochemicals in soils – the role of soil and rhizosphere microorganisms. In: Macías FA, Galindo JCG, Molinillo JMG, Cuttler HG (eds) Allelopathy chemistry and mode of action of allelochemics. CRC Press, Boca Raton, FL, pp 57–76

Blum U (2006) Allelopathy: a soil system perspective. In: Reigosa MJ, Pedrol N, González L (eds) Allelopathy a physiological process with ecological implications. Springer, Dordrecht, The Netherlands, pp 299–340

Blum U (2007) Can data derived from field and laboratory bioassays establish the existence of allelopathic interactions in nature? In: Fujii Y, Hiradate S (eds) Allelopathy: new concepts and methodology. Science Publishers, Enfield, NH, pp 31–38

Blum U, Dalton BR (1985) Effects of ferulic acid, an allelopathic compound, on leaf expansion of cucumber seedlings grown in nutrient culture. J Chem Ecol 11:279–301

Blum U, Dalton BR, Shann JR (1985a) Effects of various mixtures of ferulic acid and some of its microbial metabolic products on cucumber leaf expansion and dry matter in nutrient culture. J Chem Ecol 11:619–641

Blum U, Dalton BR, Shann JR (1985b) Effects of ferulic and *p*-coumaric acids in nutrient culture on cucumber leaf expansion as influenced by pH. J Chem Ecol 11:1567–1582

Blum U, Gerig TM (2005) Relationships between phenolic acid concentrations, transpiration, water utilization, leaf area expansion, and uptake of phenolic acids: nutrient culture studies. J Chem Ecol 31:1907–1932

Blum U, Gerig TM, Weed SB (1989) Effects of mixtures of phenolic acids on leaf area expansion of cucumber seedlings grown in different pH Portsmouth A_1 soil materials. J Chem Ecol 15: 2413–2423

Blum U, Gerig TM, Worsham AD, Holappa LD, King LD (1992) Allelopathic activity in wheat-conventional and wheat-no-till soils: development of soil extract bioassays. J Chem Ecol 18:2191–2221

Blum U, Gerig TM, Worsham AD, King LD (1993) Modification of allelopathic effects of *p*-coumaric acid on morning-glory seedling biomass by glucose, methionine, and nitrate. J Chem Ecol 19:2791–2811

Blum U, Rebbeck J (1989) The inhibition and recovery of cucumber roots given multiple treatments of ferulic acid in nutrient culture. J Chem Ecol 15:917–928

Blum U, Shafer SR (1988) Microbial populations and phenolic acids in soils. Soil Biol Biochem 20:793–800

Blum U, Staman KL, Flint LJ, Shafer SR (2000) Induction and/or selection of phenolic acids-utilizing bulk-soil and rhizosphere bacteria and their influence on phenolic acid phytotoxicity. J Chem Ecol 26:2059–2078

Blum U, Wentworth TR, Klein K, Worsham AD, King LD, Gerig TM, Lyu S-W (1991) Phenolic acid content of soils from wheat-no till, wheat-conventional till, and fallow-conventional till soybean cropping systems. J Chem Ecol 17:1045–1068

Box JD (1983) Investigation of the Folin & Ciocalteau phenol reagent for the determination of polyphenolic substances natural in waters. Water Res 17:511–525

Bradow JM (1991) Relationships between chemical structure and inhibitory activity of C_6 through C_9 volatiles emitted by plant residues. J Chem Ecol 17:2193–2212

Bradow JM, Connick WJ (1990) Volatile seed germination inhibitors from plant residues. J Chem Ecol 16:645–666

Buttery RG, Xu C-J, Ling LC (1985) Volatile components of wheat leaves (and stems): Possible insect attractants. J Agric Food Chem 33:115–117

Cheremisinoff PN, Ellerbusch F (1978) Carbon adsoprtion handbook. Ann Arbor Science Publishers Inc, Ann Arbor, MI

Dalton BR (1993) Extraction and behavior of plant phenolic acids in soils. North Carolina State University Thesis, Raleigh, NC

Dalton BR (1999) The occurrence and behavior of plant phenolic acids in soil environments and their potential involvement in allelochemical interference interactions: methodological limitations in establishing conclusive proof of allelopathy. In: Inderjit, Daskshini KMM, Foy CL

(eds) Principles and practices in plant ecology: allelochemical interactions. CRC Press, Boca Raton, FL, pp 57–74
Dao TH (1987) Sorption and mineralization of plant phenolic acids in soil. In: Waller GR (ed) Allelochemicals: role in agriculture and forestry. ACS symposium series, vol 330. American Chemical Society, Washington, DC, pp 358–370
Debyle NV, Hennes RW, Hart GE (1988) Evaluation of ceramic cups for determining soil solution chemistry. Soil Sci 146:30–36
DeFrank J, Putnam AR (1978) Weed and crop response to allelopathic crop residues. North Cent Weed Control Conf Proc 33:44
DeFrank J, Putnam AR (1985) Screening procedures to identify soil-born actinomycetes that can produce herbicidal compounds. Weed Sci 33:271–274
Duke SO, Cedergreen N, Velini ED, Belz RG (2006) Hormesis: is it an important factor in herbicide use and allelopathy? Outlook Pest Manag 16:29–33
Einhellig FA, Schon MK, Rasmussen JA (1982) Synergistic effects of four cinnamic acid compounds on grain sorghum. J Plant Growth Reg 1:251–258
Flaig W (1971) Organic compounds in soil. Soil Sci 111:19–33
Fry CF (1988) The growing plant cell wall: chemical and metabolic analysis. Longman Scientific and Technical, Harlow, Essex
Gerig TM, Blum U (1991) Effects of mixtures of four phenolic acids on leaf area expansion of cucumber seedlings grown in Portsmouth B1 soil materials. J Chem Ecol 17:29–40
Gerig TM, Blum U (1993) Modification of an inhibition curve to account for effects of a second compound. J Chem Ecol 19:2783–2790
Giesler R, Lundström U, Grip H (1996) Comparison of soil solution chemistry assessment using zero-tension lysimeters or centrifugation. Eur J Soil Sci 47:395–405
Glass ADM (1973) Influence of phenolic acids on ion uptake. I. Inhibition of phosphate uptake. Plant Physiol 51:1037–1041
Glass ADM (1974) Influence of phenolic acids on ion uptake. III. Inhibition of potassium uptake. J Exp Bot 25:1104–1113
Glass ADM (1975) Inhibition of phosphate uptake in barley roots by hydroxy-benzoic acids. Phytochemistry 14:2117–2130
Glass ADM, Dunlop J (1974) Influence of phenolic acids on ion uptake. IV. Depolarization of membrane potentials. Plant Physiol 54:855–858
Grover BL, Lamborn RE (1970) Preparation of porous ceramic cups to be used for soil extraction of soil water having low solute concentrations. Soil Sci Soc Am Proc 34:706–708
Harborne JB (1982) Introduction to ecological biochemistry. Academic Press, London
Heckman JR, Strick JE (1996) Teaching plant-soil relationships with color images of rhizosphere pH. J Nat Resour Life Sci Educ 25:13–17
Heisey RM, DeFrank J, Putnam AR (1985) A survey of soil microorganisms for herbicidal activity. In: Thompson AC (ed) The chemistry of allelopathy biochemical interaction among plants. ACS symposium series, vol 268. American Chemical Society, Washington, DC, pp 337–349
Heisey RM, Mishra SK, Putnam AR, Miller JR, Whitenack CJ, Keller JE, Huang J (1988) Production of herbicidal and insecticidal metabolites by soil microorganisms. In: Cutler HG (ed) Biologically active natural products – potential use in agriculture. ACS symposium series, vol 380. American Chemical Society, Washington, DC, pp 65–78
Huang PM, Wang TSC, Wang MK, Wu MH, Hsu NW (1977) Retention of phenolic acids by noncrystalline hydroxy-aluminum and -iron compounds and clay minerals in soil. Soil Sci 123:213–219
Huang PM, Wang MC, Wang MK (1999) Catalytic transformation of phenolic compounds in the soil. In: Inderjit, Dakshini KM, Foy CL (eds) Principles and practices in plant ecology: allelochemical interactions. CRC Press, Boca Raton, FL, pp 287–306
Hughes S, Reynolds B (1988) Cation exchange properties of porous ceramic cups: implications for field use. Plant Soil 109:141–144

Jardine PM, Wilson GV, Luxmore RJ (1990) Unsaturated solute transport through forest soil during rain storm events. Geoderma 46:103–118

Jardine PM, Wilson GV, Luxmore RJ, McCarthy JF (1989) Transport of inorganic and natural organic tracers through an isolated pedon in forest watershed. Soil Sci Soc Am J 53:317–323

Klein K, Blum U (1990) Inhibition of cucumber leaf expansion by ferulic acid in split-root experiments. J Chem Ecol 16:455–463

Laidler KJ (1998) To light such a candle: chapters in the history of science and technology. Oxford University Press, Oxford

Lavelle P, Spain AV (2001) Soil ecology. Kluwer Academic Publishers, Dordrecht, The Netherlands

Lehman ME, Blum U (1999) Evaluation of ferulic acid uptake as a measurement of allelochemical dose: effective concentration. J Chem Ecol 25:2585–2600

Lehman ME, Blum U, Gerig TM (1994) Simultaneous effects of ferulic and *p*-coumaric acids on cucumber leaf expansion in split-root experiments. J Chem Ecol 20:1773–1782

Liebl RA, Worsham AD (1983) Inhibition of morning-glory (*Ipomoea lacunosa* L.) and certain other weed species by phytotoxic components of wheat (*Triticum aestivum* L.) straw. J Chem Ecol 9:1027–1043

Loi RX, Solar MC, Weidenhamer JD (2008) Solid-phase microextraction method for in vivo measurements of allelochemical uptake. J Chem Ecol 34:70–75

Lynch JM (1977) Phytotoxicity of acetic acid produced in the anaerobic decomposition of wheat straw. J Appl Bact 42:81–87

Lyu S-W, Blum U (1990) Effects of ferulic acid, an allelopathic compound, on net P, K, and water uptake by cucumber seedlings in a split-root system. J Chem Ecol 16:2429–2439

Macías FA, Oliveros-Bastidas A, Marin D, Castellano D, Simonet AM, Molinillo JMG (2005) Degradation studies on benzoxazinoids. Soil degradation dynamics of (2R)-2-O-ß-D-glucopyranosyl-4-hydroxy-(2H)-1,4-benzoxazin-3-(4H)-one (DIBOA-Glc) and its degradation products, phytotoxic allelochemicals from gramineae. J Agric Food Chem 53:554–561

Mahall BE, Callaway RM (1992) Root communication mechanisms and intracommunity distribution of two mojave desert shrubs. Ecology 73:2145–2151

Marschner H, Römheld V (1983) In vivo measurements of root-induced pH changes at soil-root interface: effects of plant species and nitrogen sources. Zeitschrift für Planzenernähr Bodenkunde 111:241–251

Mathiassen SK, Kudsk P, Mogensen BB (2006) Herbicidal Effects of soil-incorporated wheat. J Agric Food Chem 54:1058–1063

Mattson JS, Mark HB (1971) Activated carbon: surface chemistry and adsorption from solution. Marcel Dekker Inc, New York, NY

McAllister RA (1969) Observations on the folin-ciocalteu reaction. J Med Lab Technol 26:1–10

Mohney BK, Matz T, LaMoreaux J, Wilcox DS, Gimsing AL, Mayer P, Weidenhamer JD (2009) In situ silicone tube microextraction: a new method for undisturbed sampling of root exuded thiophenes from marigold (*Tagetes erecta* L.) in soil. J Chem Ecol 35:1279–1287

Neary AJ, Tomassini F (1985) Preparation of alundum/ceramic plate tension lysimeters for soil water collection. Can J Soil Sci 65:169–177

Niemeyer HM , Pesel ES, Capaja SV, Bravo HR, Franke S, Francke W (1989) Changes in hydroxamic acid levels of wheat plants induced by aphid feeding. Phytochemistry 28:447–449

Nilsson MC (1994) Separation of allelopathy and resource competition by the boreal dwarf shrub *Empetrum hermaphroditum* Hagerup. Oecologia 98:1–7

Ohno T (2001) Oxidation of phenolic acid derivatives by soil and its relevance to allelopathic activity. J Environ Qual 30:1631–1635

Ortas I, Rowell DL (2004) Effects of ammonium and nitrate on indigenous mycorrhizal infection, rhizosphere pH changes, and phosphorus uptake by Sorghum. Commun Soil Sci Plant Anal 35:1923–1944

Patrick ZA (1971) Phytotoxic substances associated with the decomposition in the soil of plant residue. Soil Sci 111:13–18

Paul EA, Clark FE (1989) Soil microbiology and biochemistry. Academic Press, San Diego, CA

Pue KJ, Blum U, Gerig TM, Shafer SR (1995) Mechanism by which noninhibitory concentrations of glucose increase inhibitory activity of *p*-coumaric acid on morning-glory seedling biomass accumulation. J Chem Ecol 21:833–847

Putnam AR, DeFrank J, Barnes JP (1983) Exploitation of allelopathy for weed control in annual and perennial cropping systems. J Chem Ecol 9:1001–1010

Rao TP, Yano K, Iijima M, Yamauchi A, Tatsumi J (2002) Regulation of rhizosphere acidification by photosynthetic activity of cowpea (*Vigna unguiculata* L Walp) seedlings. Ann Bot 89: 213–220

Rassmusen JA, Einhellig FA (1977) Synergistic inhibitory effects of *p*-coumaric acid and ferulic acid on germination and growth of grain sorghum. J Chem Ecol 3:197–205

Ridenour WM, Callaway RM (2001) The relative importance of allelopathy in interference: the effects of an invasive weed on a native bunchgrass. Oecologia 126:444–450

SAS Institute Inc (1999) SAS/STAT user's guide, Version 8. SAS Publishing, Cary, NC

Schmidt SK, Ley RE (1999) Microbial competition and soil structure limit the expression of allelopathy. In: Interjit, Dakshini KM, Foy CL (eds) Principles and practices in plant ecology: allelochemical interactions. CRC Press, Boca Raton, FL, pp 339–351

Shafer SR, Blum U (1991) Influence of phenolic acids on microbial populations in the rhizosphere of cucumber. J Chem Ecol 17:369–389

Shann JR, Blum U (1987) The utilization of exogenously supplied ferulic acid in lignin biosynthesis. Phytochemistry 26:2977–2982

Shilling DG, Liebl RA, Worsham AD (1985) Rye (Secale cereale L.) and wheat (Triticum aestivum L.) mulch: the suppression of certain broadleaved weeds and the isolation and identification of phytotoxins. In: Thompson AC (ed) The chemistry of allelopathy: biochemical interactions among plants. ACS symposium series, vol 268. American Chemical Society, Washington, DC, pp 243–271

Sinkkonen A (2001) Density-dependent chemical interference – an extension of the biological response model. J Chem Ecol 27:1513–1523

Sinkkonen A (2003) A model describing chemical interference caused by decomposing residues at different densities of growing plants. Plant Soil 250:315–322

Sinkkonen A (2007) Modeling the effects of autotoxicity on density-dependent phytotoxicity. J Theor Biol 244:218–227

Staman K, Blum U, Louws F, Robertson D (2001) Can simultaneous inhibition of seedling growth and stimulation of rhizosphere bacterial populations provide evidence for phytotoxin transfer from plant residues in the bulk soil to the rhizosphere of sensitive species? J Chem Ecol 27: 807–829

Strobel BW (2001) Influence of vegetation on low-molecular-weight carboxylic acids in soil solution – a review. Geoderma 99:169–198

Tang CS, Waiss AC (1978) Short-chain fatty acids as growth inhibitors in decomposing wheat straw. J Chem Ecol 4:225–232

Thijs H, Shann JR, Weidenhamer JD (1994) The effects of phytotoxins on competitive outcome in a model system. Ecology 75:1959–1964

Weidenhamer JD (1996) Distinguishing resource competition and chemical interference: overcoming the methodological impasse. Agron J 88:866–875

Weidenhamer JD (2005) Biometric measurements of allelochemical dynamics in the rhizosphere. J Chem Ecol 31:221–236

Weidenhamer JD (2007) New approaches to analyze allelochemicals in soil. Allelopathy J 19: 135–142

Weidenhamer JD, Hartnett DC, Romeo JT (1989) Density-dependent phytotoxicity: distinguishing resource competition and allelopathic interference in plants. J Appl Ecol 26:613–624

Willard JI, Penner D (1976) Benzoxazinones: cyclic hydroxamic acids found in plants. Residue Rev 64:67–76

Wu H, Haig T, Pratley J, Lemerle D, An M (2001) Allelochemicals in wheat (*Triticum aestivum* L.): Production and exudation of 2,4-dihydroxy-7-methoxy-1,4-benzoxazin-3-one. J Chem Ecol 27:1691–1700

Wu H, Pratley J, Lemerle D, Haig T (2000) Laboratory screening for allelopathic potential of wheat (*Triticum aestivum*) accessions against annual ryegrass (*Lolium rigidum*). Aust J Agric Res 51:259–266

Yenish JP, Worsham AD, Chilton WS (1995) Disappearance of DIBOA-glucoside, DIBO, and BOA from rye (*Secale cereale* L.) cover crop residue. Weed Sci 43:18–20

Zabowski D (1989) Limited release of soluble organics from roots during centrifugal extraction of soil solutions. Soil Sci Soc Am J 53:977–979

Zabowski D, Ugolini FC (1990) Lysimeter and centrifuge soil solution: seasonal differences between methods. Soil Sci Soc Am J 54:1130–1135

Author Index

U. Blum, *Plant–Plant Allelopathic Interactions*, DOI 10.1007/978-94-007-0683-5,

Subject Index

T

U

V

W

FSC
www.fsc.org
MIX
Papier | Fördert
gute Waldnutzung
FSC® C083411

Kolibri 360 GmbH
Ferdinand-Jühlke-Straße 7
99095 Erfurt, Deutschland
produktsicherheit@kolibri360.de